ENTREPRENEURSHIP DEVELOPMENT

Dr. R.K. Singal
Sruti Singal

ENTREPRENEURSHIP DEVELOPMENT

by :

Dr. R.K. SINGAL
Professor and Consultant

SRUTI SINGAL
M.B.A
Manpower Consultant

S.K. KATARIA & SONS®
Publisher of Engineering Books
4885/109, Prakash Mahal, Dr. Subhash Bhargav Lane,
Opposite Delhi Medical Association, Daryaganj, New Delhi-110002 (INDIA)
Phone: +91-11-43551243; *Mobile:* +91-9871775858, 7678425209
e-mail: katariabook@gmail.com, katariabooks@yahoo.com
Website: www.skkatariaandsons.com
(e-books, subscriptions and bulk buying also available on our website)

Published by:

S.K. Kataria & Sons®

4885/109, Prakash Mahal, Dr. Subhash Bhargav Lane,
Opposite Delhi Medical Association, Daryaganj, New Delhi-110002 (INDIA)
Phone: +91-11-43551243; *Mobile:* +91-9871775858, 7678425209
e-mail: katariabook@gmail.com, katariabooks@yahoo.com
Website: www.skkatariaandsons.com

Send all correspondence to : **M/s S.K. Kataria & Sons,** New Delhi
e-mail: katariabook@gmail.com

First Edition: August, 2011; Second Edition: July, 2012; Third Edition: September, 2013
Fourth Edition: January, 2021, Reprint: 2024, Fifth Edition: 2026

Laser Typesetting
PLUS COMPUTERS, Shahdara, Delhi-93

Printed at
New A.S. Printing Press, Delhi-06

PREFACE

The future belongs to entrepreneurial socities. The Centre and State Governments have recognised the role of Small Scale Industries and startup for economic growth of the country. Entrepreneurship Development is one of the core competencies of technical human resource. Motivating entrepreneurs to start their own small scale business or industries has assumed high priority in the industrial policy of India.

Keeping in view the priorities of Government policy, courses in Entrepreneurship Development have been developed by A.K. Technical University for the engineering students of Abdul Kalam Technical University, Lucknow and Uttrakhand Technical University, Dehradun. The present book has been compiled to help the student community in this subject. It covers the syllabus fully and comprehensively. Every effort has been made to describe the matter in clear and simple way.

The authors are grateful to the publisher for bringing out this book in a very short time. The book has been prepared keeping in mind the difficulties and requirements of the students. Any constructive criticism will be welcome with gratitude.

Authors

In Loving Memory of.....

MRIDUL: A Noble Soul

He lived a life , very simple and straight,
Walked stylishly through Heaven's gate.

Led a royal life without prejudice or fear,
He was a role model for all dear and near.

No lust for position, money or career,
Lived on his own terms like a warrior.

Never measured achievements through money,
Lived to his name, always distributing honey.

"Give more to life than take from it,"
He consumed his life without a tilt.

For him I was his father, deeply allied,
But he was my friend, philosopher, guide.

Talented to provide short solutions and tips,
No queries left without replies with wits.

Had all the twenty six qualities of a gentleman.
Advocated by Lord Krishna for a worthy man.

Wish he had continued a little more,
And lighted the Earth like a meteor.

RK Singal
A Proud Father

Note: These are unedited verses as Mridul is not available for editing.

CONTENTS

PART–I: ENTREPRENEUR AND SMALL SCALE INDUSTRIES 1–48

1. ENTREPRENEUR ... 3–20
1.1 Introduction ... 3
1.2 Meaning of Entrepreneur and Entrepreneurship ... 4
1.2.1 Concept of Entrepreneur ... 4
1.2.2 Definition of Entrepreneur and Entrepreneurship ... 5
1.3 Need for Entrepreneurship Development ... 6
1.3.1 National Need ... 6
1.3.2 Need for Technical Entrepreneurs ... 7
1.4 Scope of Entrepreneurship ... 7
1.5 Rewards and Penalties for an Entrepreneur ... 8
1.6 Role of an Entrepreneur in Socio-economic Development ... 9
1.7 Qualities of an Entrepreneur ... 10
1.8 Working Profile of an Entrepreneur ... 12
1.9 Types of Entrepreneurs ... 13
1.9.1 Types of Entrepreneurs by Clarence Danhof ... 13
1.9.2 Other Types of Entrepreneurs ... 15
1.10 Functions of an Entrepreneur ... 16
1.11 Barriers to Entrepreneurship ... 17
1.12 Entrepreneur and Manager ... 17
1.13 Theories of Entrepreneurship ... 18
1.13.1 Max Weber's Theory of Social Change ... 18
1.13.2 Hoselitz Sociological Theory ... 18
1.13.3 Trait Theory of Entrepreneurship ... 18
1.13.4 Economic Theory of Entrepreneurship ... 19
1.13.5 Schumpeter's Innovation Theory ... 19
1.13.6 McClelland's Psychological Theory ... 19
1.13.7 Theory of Social Behaviour ... 19
Question Bank ... 20

2. SMALL SCALE INDUSTRIES ... 21–48
2.1 Introduction ... 21
2.2 Definition and Types of Small Scale Industries ... 22
2.2.1 New Classification of Types of Small Scale Industries ... 24
2.2.2 Registration Process ... 24
2.3 Growth of Small Scale Industries in Developing Countries ... 24
2.3.1 Need of Small Scale Industries ... 25
2.3.2 Aim and Objectives ... 25
2.3.3 Scope ... 26
2.3.4 Role of Government ... 26
2.4 Relation Between Small and Large Units ... 26
2.5 Role of Small Scale Industries in the National Economy ... 27
2.5.1 Economic Parameters ... 27
2.5.2 Performance of Small Scale Enterprises ... 27
2.6 Characteristics of Various Types of Small Scale Industries ... 28
2.7 Types of Small Scale Industries ... 30

2.8 Reasons of Sickness in Small Industries 30
2.8.1 Internal Causes 31
2.8.2 External Causes 32
2.8.3 Problems Faced by Small Enterprizes 33
2.9 Rehabilitation of Sick Units 36
2.10 Government Policy for Small-Scale Industries 38
2.10.1 Protective Measures 38
2.10.2 Promotional Measures 38
2.10.3 Institutional Measures 39
2.11 Highlights of Industrial and Licencing Policy of India 39
2.11.1 Background 39
2.11.2 Classification of Industries 40
2.11.3 Industrial Policy Statements After Independence 40
2.12 New Small Enterprize Policy, 1991 After Liberalization of Economy 42
2.12.1 Comprehensive Policy Package, 2000 43
2.12.2 Micro, Small and Medium Enterprizes Development Act, 2006 43
2.13 Special Incentives for Small-Scale Sector 43
2.13.1 Special Incentives and Subsidy are given to Small Scale Sector to Achieve the Following 43
2.13.2 Different Incentive Schemes by Central and State Governments 44
2.14 The Economic Package for the MSMEs Announced on 13th May, 2020 44
2.15 Stages in Starting a Small-Scale Industry 45
2.15.1 Time Schedule Plan for Starting a Business 46
2.15.2 Agencies to be Contacted for Project Implementation 46
Question Bank 48

PART–II: BUSINESS PLANNING **49–96**

3. PROJECT IDENTIFICATION AND SELECTION 51–70
3.1 Introduction 51
3.2 Project Identification 51
3.2.1 The Process of Selection 52
3.2.2 Exploring Opportunities 52
3.2.3 Sources for Identifying Business Opportunities 52
3.2.4 Criteria for Selecting a Particular Project 53
3.2.5 Importance of Project Identification 54
3.2.6 Constraints in Implementation of Projects 54
3.3 Project Formulation 55
3.3.1 Elements of Project Formulation 55
3.3.2 Feasibility Analysis 56
3.3.3 Techno-economic Study 57
3.3.4 Project Design and Network Analysis 58
3.3.5 Input Analysis 58
3.3.6 Financial Analysis 58
3.3.7 Cost Benefit Analysis 58
3.3.8 Precautions 59
3.4 Financing of Enterprize 59
3.4.1 Need for Finance 59
3.4.2 Sources of Finance 60
3.4.3 Capital Structure 60
3.4.4 Raising of Loans 61

3.5 Field-study and Collection of Information 62
3.5.1 Field Study 62
3.5.2 Data Collection 62
3.6 Theories of Industrial Location 64
3.6.1 Factor Affecting the Site Selection of Plant 64
3.6.2 Steps in Enterprize Locations 66
3.6.3 Analysis of Site Selection 68
Question Bank 70

4. PROJECT REPORT PREPARATION 71–96
4.1 Introduction 71
4.2 Preliminary Project Report 71
4.2.1 Advantages of Preliminary Project Report 72
4.2.2 Preparation of Preliminary Project Report 72
4.3. Techno-Economic Feasibility Study 76
4.4 Detailed Project Report (DPR) 79
4.4.1 Contents of Detailed Project Report 79
4.4.2 Preparation of Project Report 80
4.5 Project Viability 82
4.6 Project Appraisal 85
4.7 Demand Analysis 87
4.7.1 Material Balance and Output Method 88
4.7.2 Benefit Cost Analysis 89
4.7.3 Discounted Cash Flow Method 90
4.7.4 Internal Rate of Return Method 92
4.7.5 Net Present Value Method (NPV) 92
Question Bank 94

PART–III: ACCOUNTANCY AND MANAGEMENT OF SMALL BUSINESS 97–164

5. ACCOUNTANCY 99–124
5.1 Introduction 99
5.2 Purpose of Accountancy 99
5.3 Requirements 99
5.4 Procedure 100
5.5 Office of Accountant 101
5.5.1 Qualities of an Accountant 101
5.5.2 Books to be Maintained 101
5.5.3 Stationery for Accounts 101
5.6 Classification of Accounts 101
5.6.1 Double Entry Book Keeping 102
5.6.2 Groups of Accounts 102
5.7 Assessment of Economic Viability 103
5.8 Decision Making 106
5.8.1 Characteristics of Decision-Making 106
5.8.2 Types of Decision 106
5.8.3 Procedure of Decision-Making 107
5.9 Expected Costs 108
5.9.1 Functions of Cost Estimation 108
5.9.2 Cost Estimation Procedure 109

5.9.3 Elements of Cost and Ladder of Cost 110
5.9.4 Depreciation 112
5.9.5 Overhead Expenses 116
5.9.6 Control of Operating Costs 118
5.9.7 Control of Capital Cost 119
5.10 Preparation of Financial Reports 120
5.10.1 Trading Account 121
5.10.2 Profit and Loss Account 121
5.10.3 Balance Sheet 121
5.11 Study of Accounts and Stores 122
5.11.1 Audit of Accounts 122
5.11.2 Stock-taking of Stores 122
Question Bank 124

6. MANAGERIAL ASPECTS OF SMALL BUSINESS 125–164

6.1 Introduction 125
6.2 Definition of Management 125
6.2.1 Principles of Management 125
6.2.2 Functions of Management 126
6.2.3 Necessity of Management 128
6.2.4 Symptoms of Bad Management 128
6.3 Planning and Production Control 128
6.3.1 Advantages of Effective Prodcution Planning and Control 129
6.3.2 Functions of Production Planning and Control 130
6.4 Quality Control 135
6.4.1 Inspection 135
6.4.2 Statistical Quality Control 138
6.4.3 Total Quality Management 140
6.5 Marketing 142
6.5.1 Marketing Activities 142
6.6 Industrial Relations 147
6.6.1 Industrial Disputes 149
6.6.2 Agitation, Strikes, Lockouts, Picketing and Gherao 150
6.6.3 Settlement of Industrial Disputes 151
6.7 Sales Promotion 152
6.8 Material Purchase 153
6.8.1 Purchase Documents 155
6.9 Advertisement 157
6.9.1 Functions of Advertising 157
6.9.2 Advantages to Entrepreneur 158
6.9.3 Advantages to Customers 158
6.9.4 Requirement of a Good Advertisement 158
6.9.5 Media for Advertisement 158
6.10 Wages and Incentives 159
6.10.1 Wages 159
6.10.2 Incentives 160
6.10.3 Promotions as Motivator 160
6.10.4 Job Satisfaction as Motivator 161
6.11 Inventory Control 161
6.11.1 Need for Inventory Management 161

6.11.2 Advantage of Holding Inventories 162
6.11.3 Models of Inventory Management 162
Question Bank 163

PART–IV: PROJECT PLANNING AND CONTROL 165–196

7. PROJECT PLANNING 167–183
7.1 Introduction 167
7.2 Project Planning 167
7.2.1 Planning 167
7.2.2 Organizing 168
7.2.3 Executing 168
7.2.4 Directing and Controlling 168
7.2.5 Closing 168
7.3 The Financial Functions 169
7.4 Capital Expenditures of an Entreprize 170
7.4.1 Physical Resources 170
7.4.2 Technological Resources 171
7.4.3 Manpower or Human Resources 171
7.4.4 Financial Resources 171
7.5 Cost of Capital 172
7.5.1 Components of Cost of Capital 172
7.5.2 Factors Affecting Cost of Capital 172
7.6 Cost of Capital and Project Planning and Control 173
7.6.1 Financing of a Project 173
7.6.2 Types of Financing Options 173
7.6.3 Procurement of Assets 174
7.7 Economic Evaluation 175
7.8 Risk Analysis 175
7.8.1 Factors Leading to Uncertainty 176
7.8.2 Types of Project Risks 176
7.8.3 Techniques of Risk Analysis 177
7.9 Public Enterprises 179
7.9.1 Characteristics 179
7.9.2 Capital Expenditure 180
7.9.3 Policies 181
7.9.4 Practices 181
7.9.5 Indian Experiment 182
7.10 SWOT Analysis 182
Question Bank 183

8. PROJECT CONTROL 184–196
8.1 Introduction 184
8.2 Profit Planning and Programming 184
8.2.1 Advantages of Profit Planning 185
8.8.2 Profit Programming 185
8.3 Planning Cash Flows 186
8.3.1 Project Life 187
8.3.2 Basic Principles 187
8.3.3 Post-tax Principle 188
8.3.4 Cash Flow Calculations 188

8.4 Capital Expenditure and Operations 189
8.4.1 Management of Working Capital 189
8.5 Control of Financial Flows 192
8.5.1 Construction of Financial Flow Graph 192
8.5.2 Observations 193
8.6 Controlling 194
8.6.1 Components of a Control System 194
8.6.2 Principles of Control Process 194
8.6.3 Applications 195
8.7 Communication 195
Question Bank 196

PART–V: LEGAL OBLIGATIONS AND ENTREPRENEURIAL SUPPORT SYSTEM 197–249

9. LEGAL OBLIGATIONS 199–226
9.1 Introduction 199
9.2 Laws for Entrepreneur 199
9.3 Formation of Business Organization 200
9.3.1 Selection of Business Organization 200
9.3.2 Partnership Laws 201
9.4 Business Ownership 201
9.4.1 Sole Proprietorship 202
9.4.2 Partnership Firm 203
9.4.3 Joint Hindu Family Business 205
9.4.4 Joint Stock Company 206
9.4.5 Cooperative Organization (or Society) 208
9.4.6 Comparison of Different Forms of Business Ownership 211
9.5 Industrial Acts 211
9.5.1 Factory Act, 1948 211
9.5.2 The Indian Boiler Act, 1923 214
9.5.3 Electricity Safety Act, 1910 214
9.5.4 The Water (Prevention and Control of Pollution) Act, 1974 214
9.5.5 The Air (Prevention and Control of Pollution) Act, 1981 214
9.6 Principles of Taxation 215
9.7 Income Tax 215
9.7.1. Computation of Business Income 216
9.7.2 Deductions 217
9.7.3 Tax-payer's Obligation 218
9.8 Excise Duty 218
9.9 Sales Tax 219
9.10 Goods and Service Tax (GST) 220
9.10.1 Definition and Scope 220
9.10.2 Registration and Threshold 221
9.10.3 Composition Scheme 221
9.11 Taxation Benefits to small-scale Industry 222
9.12 Labour Laws 223
9.12.1 Workman Compensation Act, 1923 224
9.12.2 Payment fo Wages Act, 1936 224
9.12.3 Minimum Wages Act, 1948 224

9.12.4 Employees State Insurance Act, 1948 ... 224
9.12.5 Employees' Provident Fund Act, 1952 ... 225
9.12.6 Industrial Dispute Act, 1947 ... 225
9.13 Incentives and Tax Benefits for Export ... 225
9.13.1 Export Incentives ... 225
9.13.2 Tax Benefits ... 225
Question Bank ... 226

10. ENTREPRENEURIAL SUPPORT SYSTEM ... 227–249

10.1 Introduction ... 227
10.2 Sources of Information ... 227
10.3 Application Forms ... 229
10.4 District Industry Centres (DICs) ... 230
10.5 Commerical Banks ... 232
10.5.1 State Bank of India (SBI) ... 233
10.5.2 Punjab National Bank (PNB) ... 233
10.5.3 Indian Bank ... 234
10.5.4 Bank of Baroda ... 234
10.5.5 Bank of India ... 234
10.5.6 Canara Bank ... 234
10.5.7 Grindlays Bank Limited ... 235
10.5.8 Syndicate Bank ... 235
10.5.9 Other Banks ... 235
10.5.10 New Entrepreneur Development Agency, 1986 ... 235
10.6 Small Industries Service Institutes (SISIs) ... 236
10.7 National Small Industries Corporation (NSIC) ... 236
10.8 Small Industries Development Bank of India (SIDBI) ... 237
10.9 Important Institutions at State Level ... 238
10.10 National Bank for Agriculture and Rural Development (NABARD) ... 239
10.11 Special Financial Services Provided by Institutes ... 241
10.11.1 Term Lending ... 241
10.11.2 Underwriting of Capital Issues ... 242
10.12 Industrial Finance Corporation of India (IFCI) ... 243
10.12.1 Incorporation and Purpose ... 243
10.12.2 Management ... 243
10.12.3 Project Finance ... 244
10.12.4 Types of Activities ... 244
10.12.5 Functions and Policies ... 244
10.12.6 Sources of Funds ... 244
10.12.7 New Promotional Schemes ... 244
10.13 The Industrial Development Bank of India (IDBI) ... 245
10.13.1 Incorporation and Purpose ... 245
10.13.2 Objectives and Functions ... 245
10.13.3 Special Schemes ... 245
10.13.4 Sources of Funds ... 246
10.14 State Financial Corporations (SFCs) ... 246
10.14.1 Incorporation and Purpose ... 246
10.14.2 Special Schemes ... 247
10.14.3 Objectives and Functions ... 247
10.14.4 Package of Assistance ... 247
10.14.5 Eligible Concerns for Financial Assistance ... 247

10.15 Unit Trust of India (UTI) 248
10.15.1 Incorporation and Purpose 248
10.15.2 Sources of Funds 248
10.15.3 Assistance Schemes 248
Question Bank 248

ANNEXURES 251–310

Annexure-I : Factory Act, 1948 253
Annexure-II : Indian Boiler Act, 1923 259
Annexure-III : Electricity Safety Act, 1910 262
Annexure-IV : The Water (Prevention & Control of Pollution) Act, 1974 267
Annexure-V : The Air (Prevention & Control of Pollution) Act, 1981 271
Annexure-VI : The Indian Partnership Act, 1932 277
Annexure-VII : Workman's Compensation Act, 1923 280
Annexure-VIII : Payment of Wages Act, 1936 285
Annexure-IX : Minimum Wages Act, 1948 288
Annexure-X : Employee's State Insurance Act, 1948 292
Annexure-XI : Employee's Provident Fund Act, 1952 296
Annexure-XII : Industrial Dispute Act, 1947 298
Annexure-XIII : List of Industries Reserved for the Public Sector 304
Annexure-XIV : List of Agencies 306
Annexure-XV : The Consumer Protection Act, 1986 308
References Books 310

UNIVERSITY PAPERS WITH SOLUTION HINTS 311–320

INDEX 321–325

PART – I

Entrepreneur and Small Scale Industries

Chapter

1

Entrepreneur

1.1. INTRODUCTION

The word 'Entrepreneurship' has been traced to the period of crusaders and has taken on various shades of measuring down the centuries. The term 'entrepreneur' was used for army leaders in the early 16th century France. It was applied for the first time in the 18th century to a businessman who buys and sells goods at certain prices. Entrepreneurship, however, was given a central position by Schumpeter (1961) who believed that an entrepreneur was a dynamic agent of change, that an entrepreneur was a catalyst who transformed increasingly physical, natural and human resources into correspondingly production possibilities.

Entrepreneurship can be described as a creative and innovative response to the environment. Doing new things or doing things that are already being done in a new way is therefore, a simple definition of entrepreneurship. It cannotes innovativeness and an urge to take risk in the face of uncertainties, and an intuition of seeing things in a way which afterwards proves to be true.

The future belongs to entrepreneurial societies. The importance of entrepreneurial resource as a critical input in the process of economic development has been widely recognized by the scholars and policy makers. Proper industrial growth and export of goods and services can be achieved by creating a new industrial culture based on high technology, innovation and quality products. This requires preparation of a large number of entrepreneurs with technical and professional qualifications in owing small scale industries especially in the hi-tech areas or who would work in industries as entrapreneurs with the commitment and characteristics of an entrepreneur.

A majority of students undergoing technical and professional education in India come from middle income groups families and therefore chose employment as easy-going careers with immediate and guaranteed monthly incomes. They are generally shy of becoming entrepreneurs. Presently, social recognition in India is also in favour of bureaucratic and management careers for engineers and professionals. Recently, trends are, however, changing and it is expected that in a couple of years there will be reversal in career plans of engineers and other professionals who will prefer self-employment as their first choice as in the USA and other industrialized countries. The support systems have to properly channelise the availability of funds to technical entrepreneurs and simplify procedures to ease tensions and risks of industrial enterprising.

1.2 MEANING OF ENTREPRENEUR AND ENTREPRENEURSHIP

1.2.1 Concept of Entrepreneur

An entrepreneur is a person who establishes and manages his own enterprize or business. The process of establishment and management is called entrepreneurship. Entrepreneurship is the most affective way of bridging the gap between scientific inventions and marketable product by creating new enterprises. An entrepreneur is the catalyst who brings in this change.

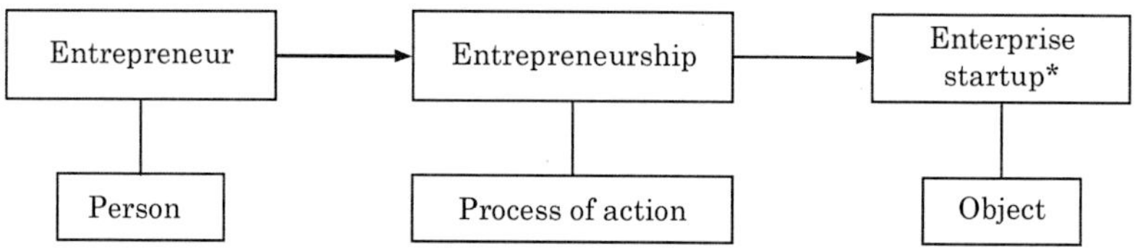

Fig. 1.1. Concept of entrepreneur.

1. Entrepreneurs are people who have the ability to see and evaluate business opportunities, to gather necessary resources to take advantage of them; and to initiate appropriate action to ensure success.
2. Entrepreneurs are action-oriented, highly motivated individuals who take risks to achieve goals.
3. An entrepreneur initiates and establishes an economic activity or enterprise for the development of wealth for the society.
4. He is certainly different from other persons in terms of his psychological and social disposition.
5. Entrepreneurs by and large, have been found to be people with a high drive and high activity level, constantly struggling to achieve something which they would call as their own accomplishment.
6. Once an entrepreneur starts his enterprize, to be successful, he would have a high need for influencing others, a low need to establish emotional relationship and a high capacity to discipline one's ownself.
7. He is inclined to speak about his past achievements and narrate them with a sense of pride.

***Startup:** A startup is a young company founded by one or more enterpreneurs to develop a unique product or service and bring it to market. By its nature, the typical startup tends to be a shoestring operations, with initial funding from the founders or their friends and families.

8. He has clear goals for the future and tends to live in the present with the involvement.
9. He studies the market situation, explores profitability in alternative lines of business, products, machinery, technology and process, finances and compares before making his final decision. This indulgence in planning may be considered as an indicator of the calculated risks taking behaviour on the part of the entrepreneur as it safeguards against subsequent difficulties which can be anticipated.
10. A successful entrepreneur continuously modifies his goals on the basis of the feedback he receives from his environment.
11. Experience with entrepreneurs indicates that their need for independence and their sense of determination are the two chief characteristics that drive them to start their own business and prefer not to be controlled by others. Analysis of the family background of the entrepreneurs indicates that in the family environment as children, their individuality was reinforced as they enjoyed ample freedom.
12. Entrepreneurs set high goals for themselves as well as their workers and drive them hard to achieve the same. While they are certainly competitive in their own orientations, they collaborate with other parties when they see such collaborations to their advantage.
13. Entrepreneurs are inclined to approach their tasks with a hope of success and tend to believe in their own capacity to influence the environment.
14. An entrepreneur is a self-centred but pushing type of person. He possesses a sense of pride in dignity of labour and exhibits readiness to put in mannual work.
15. An entrepreneur need not have all the above characteristics together. These traits can be developed in the people through psychological education.
16. A good proportion of entrepreneurs emerge from families with the entrepreneurial background or business traditions as their parents tend to socialize them towards self-reliance and independence which move them towards entrepreneurship. But entrepreneurship can be planned and developed through motivational and managerial skill development programme.

1.2.2 Definition of Entrepreneur and Entrepreneurship

Entrepreneurship is the process that involves all actions an entrepreneur undertake to establish an enterprise to give reality to his ideas. It can be described as a creative and innovative response to the environment. Entrepreneur has been defined as a risk taker, a resource assembler, an organization builder, an innovator, and so on. Following are some of the definitions of entrepreneur and entrepreneurship as given by various social scientists.

(*i*) **J.B. Say :** "Entrepreneur is the co-ordinator and organiser of resources to design a business enterprise."

(*ii*) **Adam Smith :** "Entrepreneur is an individual who undertakes the formation of an organization for commercial purposes by recognizing the potential demand for goods and services and thereby acts as an economic agent and transforms demand into supply."

(*iii*) **Joseph Shumpeter :** "Entrepreneurs are innovators who use the process of shattering the status quo of the existing products and services to set new products, new services."

(*iv*) **John Kaso and Howard Stavenson :** "Entrepreneurship is the attempt to create value through recognition of business opportunity, the management of risk-taking appropriate to the opportunity and through the communicative and management skills to mobilize human, financial and material resources necessary to bring a project to fruition."

(*v*) **Peter Druker :** "Entrepreneur is one who is involved in gathering and using resources to make use of opportunities to produce results."

"Entrepreneurship is systematic innovation which consists of purposeful and organized search for changes and in a systematic analysis of the opportunities such changes might offer for economic and social innovation."

1.3 NEED FOR ENTREPRENEURSHIP DEVELOPMENT

1.3.1 National Need

Realizing the role of small scale industries in creating larger employment opportunities, distribution of ownership, diversification of management and development of rural and backward areas, the Government of India gave major thrust to the development of small scale industries and entrepreneurship in India. It was also felt that proper industrial growth and development and export of goods and services could be achieved only by improving the industrial culture in the country. This improved culture based on high technology, innovation and quality products could be brought about by involving a large number of technical and professionally qualified entrepreneurs in owning small scale industries especially in hi-tech areas.

Since independence, the Central and State Governments have set up many organizations to nurture the culture of entrepreneurship who are prepared to take up industrial vocations and self-employment ventures especially in rural and semi-urban areas.

The Small Industries Service Institutes, National Institute of Entrepreneurship and Small Business Development at Noida and Entrepreneurship Development Institute of India at Ahmedabad are some of the few organizations to develop the entrepreneurship culture among the people. District Industries Centres and Industrial Estates provide opportunities for the entrepreneurs to set up their industries. The Entrepreneurship Development Organization, Management and Technical Institutions have made considerable impact on the young generation entrepreneurs and a number of industries have been promoted by them successfully.

1.3.2 Need for Technical Entrepreneurs

There are a large number of technical and management institutions at the degree and diploma level producing large number of technical and profession personnel. The standard of our education is comparable with international standards. India has the third largest pool of technical and scientific personnel in the world. However, we are not able to utilize its full potential and on the other hand there is surplus technical manpower who are unemployed. This large pool of technical manpower can be best utilized for developing small scale industries in the hi-tech areas using improved technologies and scientific methods of production management. They can be trained to use latest management techniques to manage the projects they set up.

A technically qualified and experienced person can make a more competent entrepreneur as he acquires special knowledge of science, engineering, materials and machines, production planning and control, manufacturing technologies and management techniques for successful launching and smooth running of an industrial unit. Also through training, he develops an aptitude for objective considerations and evaluation of issues involved in the process of an enterprize.

A technical entrepreneur develops characteristics of quality consciousness, adoption of modern technology and management techniques and realization of the importance of research and innovation for productivity improvement, the absence of which can lead to industrial sickness.

1.4 SCOPE OF ENTREPRENEURSHIP

An entrepreneur initiates and establishes an enterprize. He has to take a number of decisions to set up and run a new enterprize.

1. Decision to become an entrepreneur

It is a major step which is dictated by motivational factors. The inner urge of the person depends upon the following needs.

(*i*) To prove oneself
(*ii*) To be independent
(*iii*) To do something unique
(*iv*) To utilize skills
(*v*) To acquire greater economic reward.
(*vi*) To excel.

2. Identification and selection of an opportunity

The entrepreneur starts searching for an opportunity to meet his needs. He looks around for different possibilities of business, reads about them, meets people who could give ideas and inspiration, and collects information on several opportunities.

3. Business plan formation and its implementation

He studies the feasibility and profitability of the project. After establishment of the enterprize, he must mobilize capital and human resources and translate problems into opportunities.

4. Entrepreneurial continuum

A series of entrepreneurial decisions and actions have to be taken which require sight attitude and requisite knowledge and skills for entrepreneurial continuum as shown in Fig. 1.2.

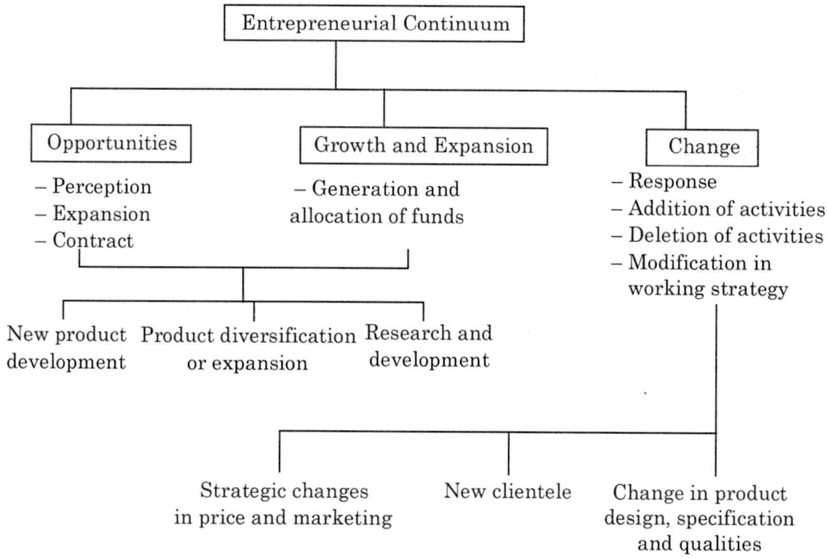

Fig. 1.2 Entrepreneurial decisions.

1.5 REWARDS AND PENALTIES FOR AN ENTREPRENEUR

I. Rewards

1. Freedom to work
2. Satisfaction of being own boss
3. Power to do things as he likes
4. Rewards of ownership and retirement assurance
5. Respect of family and friends.

II. Penalties

1. Constraints of finances, labourers, customers, suppliers and debtors control his freedom.
2. Frustration due to availability of limited capital and other resources.
3. Social and family life is affected due to hard long hours of working.
4. Frustration due to non-achievement of full objectives.
5. Risk of failure.

1.6 ROLE OF AN ENTREPRENEUR IN SOCIO-ECONOMIC DEVELOPMENT

The Government of India has realized the role of entrepreneur in the development of small scale industries and national economy. Therefore, it has given major thrust to the development of entrepreneurship and small scale industries in India.

1. Employment Generation

An entrepreneur provides employment to himself as well as is an agent for direct and indirect employment to others. He is, therefore, instrumental in solving unemployment problem of the country.

2. Entrepreneurial Culture

Proper industrial development and growth is possible only by improving the industrial culture in the society. A qualified entrepreneur is instrumental in creating industrial culture based on high technology, innovation, productivity and quality of goods and services.

3. Foreign Exchange

High quality goods and services are exported to other countries thereby earning foreign exchange for the nation.

4. Alternate Profession

An entrepreneur plays the role model for other people to start their business and shows the way for alternate profession and means of earning.

5. Competitive Price

An entrepreneur starts small business or small scale industry and can provide goods and services at competitive price, nearest to place of consumption and in time.

6. Exploitation of Consumers

An entrepreneur is able to supply goods and services at affordable prices thereby restricting the monopoly of big business houses.

7. Ancilliary Industries

An entrepreneur sets up ancilliary industry and supplies parts, components, assemblies and tooling to other industries manufacturing useful products.

8. Development of Rural and Backward Areas

An entrepreneur sets up tiny and village industries which are located in rural and backward areas. He therefore contributes to equitable development of all regions. He is doing social, justice by helping in the balanced economic growth of various regions of the country.

9. Diversification of Management

An entrepreneur sets up and manages his own enterprize and utilizes his management skills. Therefore, there is diversification of management.

10. Distribution of Ownership of Business

An entrepreneur sets up industry in his own locality and there is distribution of ownership of business and dilution in the concentration of wealth.

11. Utilization of Local Skills and Materials

An entrepreneur helps in the proper utilization of the local materials, skills, labour, capital and knowledge and therefore there is better utilization of national resources.

12. Environmental Control

Scattering of small scale industries help in reducing pollution due to concentrated discharging of affluents by large industries and thereby resulting in better environmental control.

13. Infrastructure Development

Facilities like roads, electricity, telephones, hospitals, post offices, schools, markets, housing colonies, transportation and other infrastructural development is always accompanied with the selling up of industries. Therefore, an entrepreneur contributes directly or indirectly with the overall development of a locality and region.

14. Local Employment

An entrepreneur sets up an industry and provides employment to local people. These people therefore, do not migrate to big cities thereby reducing overcrowding and overstressing of infrastructure of big cities.

The importance of entrepreneurial resource as a critical input in the process of socio-economic development has been fully recognized.

1.7 QUALITIES OF AN ENTREPRENEUR

Entrepreneurs have specific qualities and specific strength. Some of these are discussed briefly.

1. Need to Achieve

Entrepreneur has a strong desire to achieve a higher goal and fulfil his dreams. Winning is achievements.

2. Perseverence

An entrepreneur is fully absorbed in his goal and course of action. He works hard to complete his project successfully. He is not deterred by difficulties and problems.

3. Moderate Risk Taking

An entrepreneur loves moderate risk situation which is exciting but has fair chance of success. He is not a gambler or high risk taker. He selects a moderate risk and does not play wild speculative game.

4. Ability to Find and Explore Opportunity

Entrepreneur is quick to see and seize opportunities. He has innovative mind and converts difficulties into opportunities. He is also realistic. He plans and anticipates carefully to achieve his goals and converts opportunities to his advantages.

5. Analytical Ability

Entrepreneur approaches problems realistically without personal likes and dislikes. He has practical approach to business. He will select experts and not friends or relatives to help him when required. He does not take emotional attitude towards his problems.

6. Using Feedback

An entrepreneur likes to have immediate feedback of his performance. He likes quick and accurate data and information. He is stimulated by unfavourable news and applies more energy to attain his objectives.

7. Facing Uncertainty

Entrepreneur is an achievement oriented person and is optimistic even in unfamiliar situations. He has confidence in his ability to face such situation. He is not discouraged and makes the best of opportunities. He can assess the success rate closely and understand his environment. He presents a picture of boldness to face unknown situations. He will win by applying his special insight and skill.

8. Independence

An entrepreneur likes to be his own master and wants to be responsible for his own decisions. He does not want to work for others and therefore, sets up his own enterprize.

9. Flexibility

An entrepreneur has open mind. He does not hesitate in revising his decisions if situation so demands. He will evaluate his decision carefully.

10. Planning

An entrepreneur thinks in future, plans and works to make his plans successful. He will set his goals and plan to achieve within time limits.

11. Interpersonal Skills

An entrepreneur likes to work with different types of people and has the skill to deal with them. He has to make people work for him, with him and help to obtain his objectives.

12. Motivator

An entrepreneur motivates people to act. He can influence people to think his way and act accordingly.

13. Stress Taking

An entrepreneur has the capacity to work for long hours and tackle difficult problems at the same time. There will always be adverse situations in business. An entrepreneur can keep his cool under a lot of tension and make right decisions against physical and emotional stress.

14. Positive Self Concept

An entrepreneur is always positive in thinking. He is aware of his strengths and weaknesses. He always works to achieve his goals and sets standards of excellence of his works.

15. Orientation of Future

An entrepreneur shows high level of future orientation and tends to think ahead. He lives in present and does not allow the past to obsess him.

Note : No entrepreneur has all these qualities but has most of them. He should make an inventory of his qualities. Self-awareness and analysis help him to define his strengths and weaknesses. This is the first step towards entrepreneurship development.

1.8 WORKING PROFILE OF AN ENTREPRENEUR

A list of characteristics and traits provides a working profile of entrepreneurs. A list prepared by International Labour Organization of United Nations is given in Table 1.1.

TABLE 1.1. Working Profile of an Entrepreneur

Characteristics	Trait
1. Self-confidence	– Confidence
	– Independence, individuality
	– Optimism
2. Task-result oriented	– Need for achievement
	– Profit-oriented
	– Persistence, perseverence, determination
	– Hardwork, drive, energy
	– Initiative
3. Risk-taker	– Risk-taking ability
	– Likes challenges
4. Leadership	– Leadership behaviour
	– Gets along will with others
	– Responsive to suggestion, criticism
5. Originality	– Innovative, creative
	– Flexible (openness of mind)
	– Resourceful
	– Versatile, knowledgeable
6. Future-oriented	– Foresight
	– Perceptive

1.9 TYPES OF ENTREPRENEURS

Entrepreneurs are found in economic, social and cultural activities. They are found among various sections of society. Different social scientists have classified different types of entrepreneurs depending upon different criteria.

1.9.1 Types of Entrepreneurs by Clarence Danhof

Danhof has classified entrepreneurs into four categories:

1. Innovative entrepreneurs
2. Adaptive or imitative entrepreneurs
3. Fabian entrepreneurs
4. Drone entrepreneurs

1. Innovative Entrepreneurs

Innovative entrepreneur introduces new goods, inaugurates new method of production, discovers new markets and reorganizes the enterprize. He introduces something new into the economy, exploits a new source of raw material. He experiments and converts attractive possibilities into practice.

Such entrepreneurs can work and flourish only in developed economies where certain level of development has already been achieved, there are people who look forward to change and improvements. Such entrepreneurs have played important role in the rise of modern capitalism in certain developed countries.

There is a death of innovative entrepreneurs in the underdeveloped countries due to lack of resources for experimentation and innovation.

2. Adaptive or Imitative Entrepreneurs

These entrepreneurs adopt successful innovations of others from developed countries. Imitiative entrepreneurs do not innovate but make certain modifications to suit the needs of underdeveloped countries.

However, such types of entrepreneurs are required in the developing countries who can imitiate the technology, skill, etc. already available in the advanced countries, make certain modification and introduce in their developing countries. Thus these entrepreneurs face less risks of failures.

Such entrepreneurs are important for developing countries as they can perform the following functions:

(*i*) Perception of market opportunities
(*ii*) Gaining command over scarce resources
(*iii*) Purchasing inputs
(*iv*) Marketing the products
(*v*) Dealing with bureaucrats
(*vi*) Managing human relations within the enterprize
(*vii*) Managing customer and supplier relations
(*viii*) Managing finance
(*ix*) Managing production

(*x*) Acquiring and supervising setting of factory
(*xi*) Industrial engineering
(*xii*) Upgrading process and product
(*xiii*) Introducing new production techniques and products

The qualities and importance of adaptive or imitative entrepreneurs may be summed up as follows :

(*i*) Imitative entrepreneurs may not be creative but they are adaptive.
(*ii*) They are also revolutionary and important and their qualities should not be underestimated.
(*iii*) They are objective innovators and have the ability to do things which have not been done in the present enterprize.
(*iv*) Although he may not be a true innovator but he is an organiser of factors of production.
(*v*) He is a catalyst in industrialization which is very important for a developing country.
(*vi*) His actions can lead to impressive economic progress in a developing country.

3. Fabian Entrepreneur

These entrepreneurs are very cautions and sceptic in experimenting any change in their enterprises. They bring change or imitate only when it becomes perfectly sure that there would be no failure or loss.

The main qualities of such entrepreneurs are as follows :

(*i*) These entrepreneurs are neither innovative nor adaptive.
(*ii*) They are very cautious people.
(*iii*) They are rigid and follow fundamental approach.
(*iv*) They follow the foot steps of their successors.
(*v*) They are very cautious and sceptical in adapting any change.
(*vi*) They are very lazy and shy.
(*vii*) They lack will to adopt new techniques and methods.
(*viii*) Their dealings are determined by customs, religion, traditions and past practices.
(*ix*) They initiate change only when it becomes clear that they cannot survive without the change.
(*x*) They avoid risks and tend to follow their predecessors.

4. Drone Entrepreneur

These entrepreneurs do not make changes in production methods even though they may even suffer losses.

The main qualities of this type entrepreneurs are as follows :

(*i*) They are resistant to adapt and use opportunities to make changes in their products and methods of production.
(*ii*) They stick to conventional ideas and products.

(*iii*) They are laggards and continue to operate in their traditional way.

(*iv*) They are pushed out of market when product loses marketability and operations become uneconomical.

(*v*) They are very rigid and close down their business but do not accept changes.

1.9.2 Other Types of Entrepreneurs

Some more categories of entrepreneurs are given below :

1. Individual and Institutional Entrepreneurs

Individual entrepreneurs own small business and small scales industries. There are large number of individual entrepreneurs in every country. Such entrepreneurs are flexible, quick decision makers and enjoy state patronage.

A single individual can establish, operate and control only limited size of an enterprize. A group of entrepreneurs only can handle complex business and collective decision making is needed to decide line of business, amount of capital employed. Corporate sector is the symbol of institutionalized entrepreneurship.

2. Entrepreneurs by Inheritance

These entrepreneurs inherit family business by succession. The business is passed from one generation to another. They are also called second generation entrepreneurs.

First generation entrepreneurs donot have background of family business and start and build enterprize on their own innovative skills.

3. Technical Entrepreneurs

There are technically and professionally qualified persons who may enter the business to commercially exploit their invention or discoveries. They develop new goods and services out of their specialization and skills. They raise the necessary capital and employ experts in the field of finance, law, marketing and other areas of business.

4. Forced Entrepreneurs

There are persons who become entrepreneurs due to circumstances. There are many examples of this type of entrepreneurs.

(*i*) The money lenders have entered into business due to decline of money lending with the growth of banking.

(*ii*) Educated unemployed seeking self-employment mostly Indians returning from abroad.

(*iii*) Newly married bridegrooms start business with the financial support of their in-laws.

5. Basis of Motive

Entrepreneurs may be classified into following categories :

(*i*) Managing entrepreneurs whose chief goal is security

(*ii*) Innovating entrepreneurs who want excitement
(*iii*) Controlling entrepreneurs who desire power.

6. Miscellaneous Types of Entrepreneurs

Entrepreneurs may also be classified as trading entrepreneurs, agricultural entrepreneurs, industrial entrepreneurs, women entrepreneurs, rural entrepreneurs, urban entrepreneurs, etc.

1.10 FUNCTIONS OF AN ENTREPRENEUR

Entrepreneurs integrate the various factors of production to get desired results. They have to perform the following functions:

1. Sensing entrepreneurial opportunities
2. Generation of business idea
3. Converting idea into reality
4. Arranging resources
5. Supply of capital
6. Establishment of enterprize
7. Managing the enterprize
8. Growth and development

Kilby has identified thirteen functions of an entrepreneur:

1. Perceiving market opportunities
2. Gaining command over scarce resources
3. Purchasing inputs
4. Marketing of products and responding to competition
5. Dealing with public bureaucracy (concessions, licences, taxes)
6. Managing human relations within the firms.
7. Managing customer and supplier relations.
8. Managing finance
9. Managing production (control by written records, supervision, coordinating input flows into orders, maintenance).
10. Acquiring and overseeing assembly of the factory
11. Industrial engineering (minimizing inputs with given production process)
12. Updating process and product quality
13. Introducing new production techniques and products.

Kilby suggested that these functions may vary according to the size, type and setting of the enterprize and could be augmented through training and education.

1.11 BARRIERS TO ENTREPRENEURSHIP

Many entrepreneurs fail due to several barriers and problems. Vasper has identified the following barriers:

1. Lack of a viable concept which can be overcome by increased market contacts.
2. Lack of market knowledge which can be overcome by taking help of local incubator companies and market contacts, consultants.
3. Lack of technical skills which can be overcome by recruiting capable local manpower; technical education, assistance of consultants.
4. Lack of seed capital which can be arranged from supplier credit, local venture capital and bankers.
5. Lack of business know-how which can be overcome through entrepreneurial education and study of successful role models.
6. Complacency and lack of motivation can be overcome by entrepreneurial education and inspiration from successful role models.
7. Social stigma against business can be overcome by study of successful role modes.
8. Time pressures and distractions.
9. Legal contraints, regulations, red tape.
10. Monopoly and protectionism.
11. Inhibitions due to patents.

Technical education, entrepreneurship training programmes, help from consultants can reduce the effect of these barriers.

1.12 ENTREPRENEUR AND MANAGER

An entrepreneur and a manager have many common qualities of leadership. An entrepreneur can be a manager also but a manager cannot be an entrepreneur. An entrepreneur is an owner of an enterprize but a manager is an employee. The major points of difference between an entrepreneur and a manager can be compared with the help of the following table.

Difference between an entrepreneur and a manager

Points	Entrepreneur	Manager
1. Motive	The main motive of an entrepreneur is to start a varture by setting up an enterprize. He undertakes the venture for his personal gratification	The main motive of a manager is to render his expertize and services in an enterprize already set up by someone else.
2. Status	Owner of enterprize	Employee

(Cont...)

3. Risk-bearing	All risks and uncertainty involved in the success of enterprize	No risk involved
4. Rewards	Highly uncertain profit	Certain and fixed salary
5. Innovation	Inventor and change agent	Execution of plans prepared by entrepreneur
6. Qualification	Qualities and qualification of high achievement motive, originality in thinking, foresight, risk bearing ability, etc	Sound knowledge in management theory and practice

1.13 THEORIES OF ENTREPRENEURSHIP

Different theories of entrepreneurship have been proposed by different social scientists to explain various factors contributing to the emergence of entrepreneurs and entrepreneurship. Some important contributions are briefly discussed below.

1.13.1 Max Weber's Theory of Social Change

The main contents of this theory are as follows :

(*i*) Spirit of capitalism subjected to strict discipline and favourable mental attitude of society for acquisition of wealth.

(*ii*) Adventurous spirit.

(*iii*) Indians are not against the above factors and Hinduism has contributed to the development of entrepreneurship in India.

1.13.2 Hoselitz Sociological Theory

The main contents of this theory are as follows :

(*i*) An entrepreneur develops his attitude towards productivity and creative integration.

(*ii*) Entrepreneurs develop from a particular socio-economic class

(*iii*) Marwaris and parsees are dominant social classes as source of entrepreneurship in India.

1.13.3 Trait Theory of Entrepreneurship

(*i*) Entrepreneurs possess certain traits, characteristics or competencies.

(*ii*) These characteristics are required to generate new ideas and create new ventures.

(*iii*) The main traits of an entrepreneur are

– creative and innovative skills

– capacity to take risk

– organizational perserverence
– foraseeability

1.13.4 Economic Theory of Entrepreneurship

(*i*) Economic incentives are main factors for entrepreneurial activities.
(*ii*) Economic factors include :
– availability of capital
– market incentives
– institutional support
(*iii*) The main traits of an entrepreneur are
– creative and innovative skills
– capacity to take risk
(*iv*) Entrepreneurship development takes place under favourable economic conditions.

1.13.5 Schumpeter's Innovation Theory

An entrepreneur is an innovator as he brings the following :

(*i*) Introduction of new goods or new quality of goods.
(*ii*) Introduction of new methods of production.
(*iii*) Opening of new markets.
(*iv*) Finding new source of raw material
(*v*) Organization of processes based on innovation.

1.13.6 McClelland's Psychological Theory

This theory is based on psychological roots of entrepreneurship. Human motive for entrepreneurial activities drives from the following needs :

(*i*) Need for achievements : a drive to excel, advance or grow.
(*ii*) Need for power to influence others or situation.
(*iii*) Need for affilliation for friendly and close interpersonnel relationships.

1.13.7 Theory of Social Behaviour

The supply of entrepreneurship depends upon four structures found in a society :

(*i*) **Limitations structure** or certain activities (including entrepreneurship) limited by the society.
(*ii*) **Demand structure** to attract entrepreneurs for material rewards.
(*iii*) **Opportunity structure** to increase entrepreneurial activity by making available capital, management skills, engineering skills, technology, labour and markets.
(*iv*) **Labour structure**, *i.e.*, supply of competent and willing labour for industrial entrepreneurship.

Question Bank

1. Define 'entrepreneurship'. Elaborate on its scope and vital role it plays in developing economics.
2. 'Developing countries need imitative rather than innovative entrepreneurs'. Comment.
3. What are the essential characteristics of an entrepreneur ?
4. Discuss the functions of an entrepreneur.
5. Examines the role of an entrepreneur in a developing economy like India, giving appropriate examples.
6. Discuss the various reasons for entrepreneurial failure. How these problems can be reduced ?
7. How does an entrepreneur contributes to the economy and to the society ?
8. What do you understand by entrepreneurship ? Explain its need under the present employment and economic conditions in the country.
9. What is techno-entrepreneur ? Why does such a person make a more competent entrepreneur?
10. List the characteristics and traits which provides the working profile of an entrepreneur.
11. Discuss the qualities and functions of various types of entrepreneurs.

Chapter

2 Small Scale Industries

2.1 INTRODUCTION

The man appeared on this earth with certain needs which were vital for his survival. Food, water, air, shelter, temperature control, rest and relaxation, sleep, elimination of body wastes, muscular activities, sex, etc. were required to fulfill his bio-logical needs. Safety needs included avoidance of danger and harm from physical environment, animals, ailments, etc. With passage of time, he cultivated wants for relationship with others, recognition, appreciation, prestige, status, authority, dominance, etc. To meet his needs and wishes, he required materials, goods and services of others. The consumption of goods and services increased with time which resulted in the production of the same. Production systems have been created to produce goods and services efficiently and economically in an organized manner. This is called industrialization. An enterprize is an organization or a complex which carries out the production process methodically.

An enterprize is a set up to manufacture goods or provide services and earn profit. The goals of production system are fulfilled through the use of resources like materials, men, machines and money. Management is an art of getting things done through people.

An enterprize can be small-scale production of a large variety of products, medium-scale production of a limited range of products and a large-scale production of a small variety of products. The enterprize can be a small production unit employing less than 20 workers, medium unit having 21 to 1000 workers and large unit with 1001 to 10,000 people. A giant corporation may be employing more than 10,000 people.

Large scale industries require huge capital, technical skill and management abilities. These industries give large output for the capital invested and high rate of economic growth. But these industries cannot provide employment in the villages where more than 70 per cent of population lives. The sophisticated technologies adopted may make production complex and costly. They result in the emission of toxic fumes in the air and release of polluted water in the rivers and water bodies. The concentration of economic activity in the city due to large industries cause migration of village population resulting in social problems.

Entrepreneurs in small business have an advantage over large business. These small units are labour intensive and are often more efficient in the manufacture of certain goods. Small scale industries can be classified into five main types as follows :

1. Manufacturing industries producing complete articles for direct consumption and also processing industries.
2. Feeding industries specializing in certain types of components and services, e.g., castings, electroplating, welding, etc.
3. Servicing industries covering light repair shops necessary to maintain mechanical equipment.
4. Ancillary industries to large industries producing parts and components and rendering services.
5. Mining or quarrying.

2.2 DEFINITION AND TYPES OF SMALL SCALE INDUSTRIES

A small scale industry is defined in terms of size. The definition varies from one country to another and from one time to another in the same country depending upon the pattern and the stage of development. The following criteria can be used to define the size of small business.

1. Number of persons employed
2. Amount of capital invested
3. The value of annual turnover
4. Combination of number of workers and investment.

The value of annual turnover should not be used for the size of the industry as the productivity can be adversely affected.

In order to promote small scale industries in India, the Government of India had set up the Central Small Scale Industries Organization and the Small Scale Industries Board. The small scale industry was defined on the basis of investment in plant and machinery and employment criteria, i.e., maximum number of workers employed. The investment criteria has been revised from time-to-time and the employment criteria was dropped.

The small scale units have been broadly classified into seven categories.

1. Small Scale Industries

The investment in fixed assets in plant and machinery, whether held on ownership terms or on lease or by hire and purchase does not exceed ₹ 5 crore.

2. Ancillary Units

An ancillary unit must sell not less than 50% of its output to one or more parent industrial units. The limit of investment is same for ancillary units and small scale units.

3. Export Oriented Units

The unit must export atlest 50 per cent of its output by the end of three years from the date of commencement of production. The limit investment in plant and machinery is upto ₹ 5 crore.

4. Tiny Units

The investment limit for tiny industry is ₹ 25 lakh in plant and machinery. There is no restrictive condition of the location of the unit in small towns. These enterprises would be entitled to preference in land allocation, power connection, access to facilities or skills/technical upgradation. These would have also easy access to institutional finance, priority in Government purchase and relaxation in labour laws.

The micro business enterprises are included within the tiny sector and whose investment in plant and machinery does not exceed ₹ 1 lakh.

5. Village and Cottage Industries

A village industry produces any goods or renders any service with or without use of power but located in rural area. The investment should not exceed ₹ 50,000.

There is no capital investment limit of cottage industries if they have the following features:

(*i*) Capital investment is small
(*ii*) Equipment used is simple
(*iii*) Simple products in own premises
(*iv*) Organized with private resources

These are also called traditional industries.

6. Small Scale Service and Business Enterprizes

These units provide services and investment ceiling is ₹ 10 lakh. These units include marketing consultancy, advertising agencies, photocopying, desktop publishing, industrial testing laboratories, laundry and dry cleaning, x-ray clinics, tailoring, beauty parlours, etc.

7. Small Scale Industries owned and Managed by Women Entrepreneurs

An enterprize promoted by women entrepreneurs individually or jointly with share capital not less than 51 percent. The investment limit is same depending upon the category of the small scale unit. There are special concessions offered by the Government including low interest rates on loans.

The classification of small scale units is summarized in Table 2.1.

TABLE 2.1. Characteristics of Small Scale Units

S. No.	Type of Industry	Investment limit (₹)	Special features
1.	Small scale industry	5 crore	As per MSMED Act, 2006
2.	Ancillary units	5 crore	50% of output supplied to parent units
3.	Export oriented units	5 crore	Can sell 50% in domestic markets
4.	Tiny units	25 lakh	No locational restriction
5.	Village and cottage industry	50 thousand	Located in rural areas

(Contd....)

6.	Service and business units	10 lakh	No locational restriction
7.	Women enterprizes	Any of above	51% equity holding by women and managed by women.

2.2.1 New Classification of Types of Small Scale Industries

The following criteria for classification of micro, small and medium enterprises have been notified in 2020.

(*i*) *a micro enterprise*, where the investment in Plant and Machinery or Equipment does not exceed one crore rupees and turnover does not exceed five crore rupees;

(*ii*) *a small enterprise*, where the investment in Plant and Machinery or Equipment does not exceed ten crore rupees and turnover does not exceed fifty crore rupees;

(*iii*) *a medium enterprise*, where the investment in Plant and Machinery or Equipment does not exceed fifty crore rupees and turnover does not exceed two hundred and fifty crore rupees.

2.2.2 Registration Process

1. The form for registration shall be as provided in the Udyam Registration portal.
2. There will be no fee for filing Udyam Registration.
3. Aadhaar number shall be required for Udyam Registration.
4. The Aadhaar number shall be of the proprietor in the case of a proprietorship firm, of the managing partner in the case of a partnership firm and of a karta in the case of a Hindu Undivided Family(HUF).
5. In case of a Company or a Limited Liability Partnership or a Cooperative Society or a Society or a Trust, the organisation or its authorised signatory shall provide its GSTIN and PAN along with its Aadhaar number.
6. In case an enterprise is duly registered as an Udyam with PAN, any deficiency of information for previous years when it did not have PAN shall be filled up on self-declaration basis.
7. No enterprise shall file more than one Udyam Registration: Provided that any number of activities including manufacturing or service or both may be specified or added in one Udyam Registration.
8. Whoever intentionally misrepresents or attempts to suppress the self-declared facts and figures appearing in the Udyam Registration or updation process shall be liable to such penalty as specified under section 27 of the Act.

2.3 GROWTH OF SMALL SCALE INDUSTRIES IN DEVELOPING COUNTRIES

The small scale industries have the following advantages:

– Provide immediate large scale employment

- Ensure a more equitable distribution of national income
- Facilitate effective mobilization of resources such as capital and skill
- Solve problems of unplanned urbanization
- Starting small centres of industrial production all over the country.

2.3.1 Need of Small Scale Industries

Development and growth of small scale industries in developing countries like India is essential due to the following reasons.

1. Employment potential
2. Equality potential
3. Decentralization potential
4. Resource mobilization potential

1. Employment Potential

The developing countries have scarcity of capital resources and abundance of labour. The small scale units are more labour intensive than large unit and therefore employ more labour per unit of output than investment. The employment potential of small scale unit help to solve the problem of widespread unemployment in developing countries.

2. Equality Potential

Small scale industries can ensure more equitable distribution of national income and wealth. There are two reasons:

(*i*) Ownership of small-scale units is more widespread than large-scale units.

(*ii*) Decentralization and disposal of small-scale units to rural and remote areas provide more employment opportunities to the local population.

3. Decentralization Potential

Decentralization of industrial units help to tap local resource such as raw materials, idle savings, local talents. Therefore, there is improvement in the standard of living of more population.

4. Resource Mobilization Potential

The latent resources of entrepreneurship can be tapped by the growth of small enterprises only and there is emergence of entrepreneurial class of people. Future belongs to entrepreneurial societies.

2.3.2 Aim and Objectives

The aim and objectives of developing small-scale industries in a developing country in a more orderly manner can be listed by follows:

(*i*) Generation of immediate and large scale employment opportunities with relatively less investments.

(*ii*) Eradication of problems of unemployment from the country.

(*iii*) Dispersal of industries to small towns, villages and backward areas.

(*iv*) Bringing backward areas into main stream of national development.

(*v*) Balanced development of all regions.

(*vi*) Equitable distribution of national income and wealth.

(*vii*) Effective utilization of untapped resources.

(*viii*) Improvement in the living standard of all population.

2.3.3 Scope

The following types of small-scale units can be developed:

(*i*) Manufacturing units
(*ii*) Servicing and repairing units
(*iii*) Retailing shops
(*iv*) Financial enterprises
(*v*) Whole-sale business
(*vi*) Construction firms
(*vii*) Transportation, communication and other public utilities.

2.3.4 Role of Government

The Government of India has taken the following major steps to ensure growth of small-scale units:

(*i*) Assistance programmes through network of support agencies.
(*ii*) Financial assistance through special institutions.
(*iii*) Reservation for small sector for the exclusive development and insulation from unequal competition of large industrial enterprize.
(*iv*) Entrepreneurship development as part of policy and practice. More the growth of small enterprises, more will be the opportunities for entrepreneurial career and vice versa.

2.4 RELATION BETWEEN SMALL AND LARGE UNITS

In a developing economy like India there is a need for both type of enterprises, small-scale as well as large-scale. Large industries require huge capital, high technical skills and sophisticated management abilities. These industries give large output for the capital invested and high rate of economic growth. But these industries have their own shortcomings. Small scale industries are required for more balanced socio-economic development of the society. These two types of industries do not complete with each other. In fact, these industries complement to each other.

The relationship between the small and large industrial units can be seen under the following aspects.

1. Competition

Small-scale industry cannot complete with large industry under certain circumstances and for certain products. Therefore, Government of India has tried to separate the role and contribution by the two sectors. Small-scale sector has been isolated by certain Government Policies such as reservation policy for small industries and other development and assistance programmes.

2. Supplementary Role

Small scale units can fill the gaps between large scale production of certain item and other supplementary items. A small tricycle factory can flourish alongside a large cycle factory.

3. Complementary Role

Under complementary role, small units can function as ancillary units for the large units and enjoy the advantage of protected market for their products, i.e., goods and services. The subcontacting relationship between small and large industries is a glaring example of economic success of Japan and other developed and developing countries.

4. Role Model

Large units making high profits can act as role model for small units to take initiative to produce some products.

Electronic industry and some segments of automobile industry have the potential to fit into this category.

5. Service Industries

Small industries can serve the large industries by installing servicing and repair shops for the products of large units. There are examples like refrigerators, washing machines, radio and television sets, cycles and motor cycles, clocks and watches, etc.

2.5 ROLE OF SMALL SCALE INDUSTRIES IN THE NATIONAL ECONOMY

The small scale industries have played a very important role in the economic development of India. The economic development is measured by an increase in real per capita income of a person resulting in improvement in the standard of living.

2.5.1 Economic Parameters

The economic parameters to measure the contribution of small scale industries to national economy can be defined in two ways:

1. Absolute Parameters

The relevant parameters can be as follows:

(*i*) Annual increase in the number of small scale units.

(*ii*) Annual increase in the production or turn-over of small scale units.

(*iii*) Annual increase in the generation of employment.

(*iv*) Annual increase in the amount of exports.

The data can be in the form of tables or graphs.

2. Relative Parameters

The most plausible way of viewing at the role of small scale sector in economic development is by the study of its relative position in terms of country's total production, employment and exports.

2.5.2 Performance of Small Scale Enterprises

The performance of small scale sector can be viewed as follows:

1. Contribution to industrial growth.
2. Contribution to employment generation
3. Contribution to exports.

1. Industrial Growth

There is a continuous growth in the number of small scale units and their turn-over or production. The number of units increased from 109.5 lakh in 2002-03 to 128.4 lakh in 2006-07. The share of small enterprises in the country's value of output in manufacturing sector is about 39 percent. The rate annual growth of turnover is more than 12 percent.

2. Employment Generation

The small scale sector employed 263.7 lakh people in 2002-03 which rose to 312.5 lakh people in 2006-07. The total employment in the industrial sector as a whole (large, medicine and small-scale industries) increased only by 2.21 percent per annum during the period 1972-73 to 1987-88, the employment in small-scale sector grew at the rate of 5.45 percent per annum during the same period.

3. Exports

The total value of exports from small-scale sector rose from ₹ 86,013 crore in 2002-03 to ₹ 1,50,242 crore in 2005-06. The small scale sector contributes about one-third of total exports of the country.

4. Miscellaneous

Small scale industries contribute to increase in per capita income, generate immediate employment opportunities with relatively low capital investment, promotes more equitable distribution of national income, makes effective mobilization of untapped capital and human skills and leads to dispersal of manufacturing activities all over the country, leading to balanced regional development and growth of villages, small towns and backward areas. It is hoped that the relative share of small scale sector in the national production, employment and exports will increase in the years to come.

2.6 CHARACTERISTICS OF VARIOUS TYPES OF SMALL SCALE INDUSTRIES

The small scale industries are promoted and managed by one or a few people. As the level of operation is limited, the structure, organization, management and controls are altogether different than large scale industries.

The main distinguishing characteristics of small scale industries are discussed below:

1. Investment

Small industries are fairly labour intensive with comparatively smaller capital investment than the larger units. With any given investment, employment possibilities would be ten to twenty times greater in compression to large industries. The investment in plant and machinery is limited to ₹ 5 crores.

2. Organization

A small scale unit is generally a one man show. Even the small units may have proprietorship, partnership, family business, cooperative or private limited company as the organization, the activities are mainly carried out by one of the partners or directors. In practice, the others are simply inactive or sleeping partners who mainly assist in providing funds.

3. Management

The owner or promoter himself/herself is the manager also. Thus, these units are mainly managed in a personalized manner. He takes effective participation and decisions in all matters. He has knowledge of all aspects of management and operation.

4. Resources

These units use local resources such as raw materials, labour, etc., and therefore, can be located anywhere subject to availability of these resources.

5. Markets

The operation of a small scale unit is generally located catering to the demands of the locality or the region where located.

6. Factory design

The unit is designed for products reserved for small scale sector. There is no competition with large units. The factory design and layout are simple using simple plant and machinery, within the limits for small scale industries.

7. Turn-over

The gestation period is less and return on investment starts early. However volume of production is usually small.

8. Flexibility

These units are more flexible to adapt changes like introduction of new products, new methods of production, new materials, new markets and new form of organization. These are highly respective to socio-economic changes.

9. Subsidies

These units are eligible for various concessions, benefits and subsidies from the government and other agencies. The profitability of these units is sensitive to these benefits as the feasibility analysis was based taking into consideration of these benefits.

10. Starting and Operation

The small scale units are registered with the district industries centre and setting up of unit is governed by the rules of the state government. The unit has to follow the rules notified by the state government from time-to-time for its operation. However, there are not many legal formalities to start a small scale industry as compared to a large scale industry.

11. Financing

The capital requirement is small and supplied by the promoters or their relations. However, financial institutions and commercial banks are very helpful and liberal to advance loans for fixed assets and working capital.

2.7 TYPES OF SMALL SCALE INDUSTRIES

The small scale industries have been classified as per investment limit in Table 2.1, however, there are different types of industries depending upon the strength and basis. Some examples are given below:

1. Resource Based Industries

Entrepreneurs are attracted to promote an industry based on resources available in abundance. The resources may include raw materials or human skills. Some examples are:

(*i*) Agricultural products or wastes
(*ii*) Forest resources or wastes
(*iii*) Marine resources, fisheries, etc.
(*iv*) Industrial wastes and by-products
(*v*) Special human skills, artisans.

2. Demand Based Industries

Entrepreneurs promote a venture to meet good demand of a product or service locally, or to meet the gap between demand and supply. Entrepreneurs with more strength in production than marketing take advantage of the situation. Once started, they can develop capabilities to withstand demand fluctuation in the future.

3. Sub-controlled Ancillaries

These units are promoted, controlled, financial and managed by large industry. The products are sold to the promoting industry. The goods are produced at a lower cost due to benefits available to small scale units.

4. Ancillary Industry

Many large industries do not manufacture all components, parts, sub-assemblies, assemblies, toolings, etc., and develop small scale ancillary industries in the neighbourhood. There is cost advantage to manufacture there items in a small unit due to benefits and concessions available to small sector and also due to lower overhead expenses. These ancillary industries have to sell 50 percent of their products to parent industries. Some percentage can be sold to other consumers.

The parent industry may provide technology, raw materials, etc., to ancillary industry and products are manufactured as per drawings and specifications of the parent industry.

Automobile industry is usually an assembly unit which receives all the components, assemblies and sub-assemblies from ancillary units.

2.8 REASONS OF SICKNESS IN SMALL INDUSTRIES

A sick industrial unit fails to generate surplus on a continuous basis and depends on frequent infusion of external funds for its survival. A number of reason could be attributed to the sickness of small-scale industries in the country.

1. Management deficiency
2. Inadequate and timely availability of finance
3. Outdated technology, and
4. Marketing problems.

The causes of sickness can be listed as follows:

2.8.1 Internal Causes

1. Planning

(*i*) **Technical feasibility**

- Inadequate technical know-how
- Locational disadvantage
- Outdated production process.

(*ii*) **Economical liability**

- High cost of inputs
- Break-even point too high
- Uneconomical size of plant
- Under estimation of financial requirements
- Unduly large investment in fixed assets
- Over-estimates of demands.

2. Implementation

- Cost overruns resulting from delay in getting licences and sanctions
- Inadequate mobilization of finance.

3. Production

(*i*) **Production management**

- Inappropriate product mix
- Poor quality control
- Poor capacity utilization
- High cost of production
- Poor inventory management
- Inadequate maintenance and replacement
- Lack of timely and adequate modernization, etc.
- High wastage.

(*ii*) **Labour management**

- Excessively high wage structure
- Inefficient handling of labour problems
- Excessive manpower
- Poor labour productively
- Poor labour relations
- Lack of trained skilled labour or technically competent personnel.

(*iii*) **Marketing management**

- Dependence on single customer or a limited number of customers

- Single or limited number of products
- Poor sales realization
- Defective pricing policy
- Booking of large orders at fixed price in an inflationary market
- Weak market organization
- Lack of market feedback and market research
- Lack of knowledge of marketing techniques
- Unscuplous sales/purchase practices.

(*iv*) **Financial management**

- Poor resources management and financial policy
- Faulty costing
- Liberal dividend policy
- General financial indiscipline and application of funds for unauthoried purposes
- Deficiency of funds
- Over-trading
- Unfavourable gearing or keeping adverse debt-quality ratio
- Inadequate working capital
- Absence of cost estimation
- Lack of effective collection machinery.

(*v*) **Administrative management**

- Over-centralization
- Lack of professionalism
- Lack of feedback to management (Defective Management Information System)
- Lack of controls
- Lack of timely diversification
- Excessive expenditure on R&D
- Dissention within the management
- Incompetent management
- Dishonest management.

2.8.2 External Causes

(*i*) **Infrastructural bottlenecks**

- Non-availability or irregular supply of critical raw materials and other inputs
- Chronic power shortage
- Transport bottlenecks

(*ii*) **Financial bottlenecks**

- Non-availability of adequate finance

(*iii*) **Government control and policies**

- Government price control

- Fiscal duties
- Abrupt changes in government policies

(*iv*) **Market conditions**

- Market saturation
- Revolutionary technological advances rendering products absolete.

(*v*) **Extraneous factors**

- Natural calamities
- Political situation
- War
- Sympathetic strikes
- Multiplicity of labour unions.

2.8.3 Problems Faced by Small Enterprizes

2.8.3.1 Introduction

The small-scale enterprizes have good intensions and have shown good growth and overall contribution of small-scale sector in Indian economy has been very good. However, small scale industries have suffered from many problems as well. There are many and varied problems depending partly on the level of economic and social development of the country. The problems faced by small-scale sector and organized sector are almost identical. However, the organized sector is financially very strong with large resources. Therefore, it can face its problems more effectively. But the small entrepreneur is helpless in this regard. From the moment, a small entrepreneur conceives an idea to start his own unit, he has to work against heavy odds. The first step, *i.e.*, the preparation of a project report, calls for the collection of data on the marketability of the product chosen, the availability of raw materials, the manufacturing techniques involved, the choice of machinery and location. The entrepreneur is also confronted with problems when the enterprise is alive and kicking. If it is sick, another set of problems.

The entrepreneur faces two types of problems. The external problem like availability of power and other infrastructure which are beyond the control of entrepreneur. The internal problems such as organization structure, production channel, distribution channel, technical know-how, training industrial relations and inadequacy of management which also influence the smooth operation of the enterprize. Therefore, the small scale units are very vulnerable to go sick on account of various inherent problems.

The main problems faced by small enterprizes may be classified as follows:

1. Difficulties in marketing.
2. Management of new products.
3. Shortage of raw materials and power.
4. Under-utilization of capacity.
5. Problems of finance.

2.8.3.2 Difficulties in Marketing

One of the main problems faced by the small-scale units are in the field of marketing. These problems can be identified as follows:

1. Small units often do not possess any proper marketing organization. In consequence, their products compare unfavourably with the quality of the products of the large-scale industries. Therefore, they suffer from comparative disadvantages vis-a-vis large-scale units.
2. **Ancillary units** face the problem of getting delayed payment from parent units; inadequacy of technological support extended by parent units, non-adherence to quality and delivery schedules, thus, disturbing the programmes of the parent unit. There is an absence of a well-defined pricing system and regulatory laws.
3. The cost of production is usually higher and cannot complete with large scale units due to economy-of-scale.
4. Many times, the small scale sector cannot afford their own marketing organization.
5. Many small firms manufacture goods which are not of uniform quality. They do not follow quality control because of which they cannot after standardized products.
6. To raise money immediately, sometimes, the small scale unit has to sell their products at a very low price.
7. Small-scale units are facing problems of exporting their units as they have to face competition not only from medium and large industries but also multi-national companies. They cannot withstand their quality standards, technological skills and financial creditworthiness. They have limited access to markets of developed countries due to stringent requirements of quality certification such as ISO 9000.

Notes:

1. In order to save small-scale units from this competitive disadvantage, the Government of India has reserved certain items for small scale sector which has continuously grown to 824 items.
2. The National Small Industries Corporation is helping the small-scale units in obtaining the government orders and locating export markets.
3. Trade Fair Authority of India and the State Trading Corporation (STC) help the small-scale industries in organizing their sales.

2.8.3.3 Management of New Products

In most of the small-scale units, the ownership and management functions are performed by the owner of the enterprize. He may not possess the necessary skills of efficient planning, organising and controlling. Management is a complex job requiring knowledge and training which are not found in small-scale units. The other problem are as follows:

1. Old Technology

Most of the small firms are using old technologies because they cannot afford new machines and equipment. Therefore, their cost of production is high with less margin of profit as compared to large units which can employ advanced technology and innovative activities.

2. Lack of Trained Personnel

There is a lack of personnel with adequate managerial skills and qualities and such persons seek better opportunities of employment in the large-scale industries. There is, therefore, lack of entrepreneurs with knowledge and vision and also good managers and skilled personnel.

3. Low Labour Productivity

Labour productivity in small-scale sector is low due to limited salaries which can be paid. There is higher turnover and difficulty in attracting talented people.

4. Low Quality

Small-scale business concentrates on cost cutting and keeping the prices low. They do not have adequate resources to invest in quality research and maintain the industry standards. They cannot upgrade technology due to lack of expertise and funds.

2.8.3.4 Shortage of Materials and Power

The small-scale industries suffer from acute shortage of certain basic raw materials in the following aspects:

1. An absolute scarcity in the availability.
2. A poor quality of raw materials.
3. A high cost of raw materials.

The enterprizes producing items based on local raw materials do not face very serious problem. However, the units producing sophisticated goods using imported raw materials face raw materials problem due to shortage of foreign exchange.

The government has made certain arrangements for supply of certain raw materials to small scale industries. However, small units are not able to avail off this facility due to non-availability of special staff to liasion with official agencies. When these units purchase materials at high price from open market, these are not able to compete with large units due to high cost of production.

Shortage of power is another factor which leads to under-utilization of plant capacity. Most of the small-scale units cannot afford their own generating sets as back-up supply during non-supply of power.

2.8.3.5 Under Utilization of Capacity

There is gross under-utilization of installed capacity in small-scale industries. On an average 50 to 40 percent of capacity is not utilized in small-scale units. The main factors of under-utilization of capacity are given above. The main problem, however, is non-supply of sufficient power. Since, small-scale units are weak economically they have to manage within their merge means.

2.8.3.6 Problem of Finance

Every business enterprize needs sufficient funds to finance the purchase of plant and machinery, building, raw materials and to meet day-to-day expenses including wages and salaries, rent, power and water charges, administrative and selling expenses. Small-scale units are not able to raise sufficient money to provide for fixed and working capital. Shortage of funds is the most important factor which is responsible to under-utilization of capacity and led to the closure of many small-scale units.

The problem of finance in small-scale sector is mainly due to following reasons:

1. Partly due to scarcity of capital in the country as a whole.
2. Partly due to weak credit worthiness of the small entrepreneur.
3. Due to their weak economic base, these units find it difficult to take financial assistance from commercial banks and financial institutions.
4. These units have to obtain credit from the money lenders on a very high rate of interest, thereby, losing their competitiveness.

Note : Due to new policy, the availability of institutional finance is actually increasing. Still the criterion of "credit worhiness" still weighs heavily on nationalized banks. It is necessary to further liberalise the rules and practices of banking in the country.

2.9 REHABILITATION OF SICK UNITS

There are far searching socio-economic evil effects of industrial sickness needing urgent solution of rehabilitation of sick industries in India. The following are the measures for rehabilitation of sick units.

1. Identification and Detection of Sickness

Industrial sickness is a gradual process taking 5 to 7 years. The identification and detection of sickness at the incipient stage should be taken to reduce it. Appropriate yardstick should be developed for early detection of sickness.

2. Merger of Sick Units

A large number of sick units cannot be revived due to lack of resources. These units should be:

(*i*) permitted to close or liquidate, or

(*ii*) some units should be picked up for revival or rehabilitation, and

(*iii*) some units may be merged together to prevent displacement of large number of labourers.

3. Salvage of Sick Units

The problems of industrial sickness have different effects on different stake holders, and, therefore they look for different solutions.

(*i*) The entrepreneur wants to close down the sick unit if it is no more viable.

(*ii*) Financial institutions and creditors want to salvage as much as possible.

(*iii*) The labour and other employees want the unit to continue otherwise they would lose job and other benefits.

The sick unit should be saved and salvaged to take care of interests of all the stake holders.

4. Board for Industrial and Financial Reconstruction (BIFR)

The sick small-scale industries fall within the purview of BIFR and this board should be provided with expertise to deal with industrial sickness for rehabilitation. The small-scale industries are characterized by different types of problems and highly competent authority can handle sickness.

5. Rehabilitation Programme

A systematic and scientific rehabilitation programme should be developed and followed for solution of sickness in different small-scale industries. The programme should cover all the aspects of sickness as per requirement of the unit.

(*i*) Rescheduling of debts of enterprize.
(*ii*) Sanction of additional term loan for installation of new machines and equipment.
(*iii*) Enhancement of working capital limit.
(*iv*) Asistance to improve managerial efficiency.
(*v*) Assistance in the marketability of products.
(*vi*) Help to ensure adequate availability of power and raw materials.
(*vii*) Other problems faced by the affected enterprize.

Therefore, a comprehensive rehabilitation programme is the requirement.

6. Timely Implementation of Rehabilitation Programme

There is an urgent need to finalize the rehabilitation programme quickly and implement the same speedily and in full. The unit will continue to make loss and additional funds injected may also be wiped out if the need-based measures are not implemented. The rehabilitation programme must ensure that the unit has started operating above the break-even point.

7. Feedback System

There should be a proper feedback system to arrest the sickness at the incipient stage.

(*i*) Banks and financial institutions should periodically review the accounts of the small-scale industries.
(*ii*) These institutions should study the units which are becoming sick or are prone to sickness.
(*iii*) These lending institutions should provide the information on sickness to BIFR and other agencies involved in the implementation of rehabilitation programmes.
(*iv*) The rehabilitation programmes should be implemented by the agencies quickly and comprehensively.
(*v*) The Government of India and the Reserve Bank should direct the commercial banks and lending institutions to work as watchdog and effectively link of feedback system.

8. Entrepreneurship Development Programme (EDP)

The entrepreneur should be imparted necessary knowledge in various functional areas through the training programme such as EDP. An entrepreneur may have land, building, machinery, etc., but no experience in production and marketing. He must prepare the project report after proper opportunity scanning and collection of all data. He should not behave like a one-man band mainly interested to avail of subsidies, concessions and incentives from the government.

2.10 GOVERNMENT POLICY FOR SMALL-SCALE INDUSTRIES

Small-scale enterprises have been given an important place in the framework of Indian planning for both ideological and economic reasons. The development of small-scale enterprises have the following objectives.

(*i*) The generation of immediate employment opportunities with relatively low investment.

(*ii*) The promotion of more equitable distribution of national income.

(*iii*) Effective mobilization of untapped capital and human skills.

(*iv*) Dispersal of manufacturing activities all over the country.

(*v*) Growth of villages, small towns and far-flung economically lagging regions.

India has built up perhaps one of the world's most elaborate programmes for small enterprise development for providing assistance to entrepreneurs for setting up small-scale enterprises. The governments objectives and intensions were announced through Industrial Policy Resolutions (IPRs) in 1948, 1956, 1977, 1980, 1990 and 1991.

The various industrial policies announced from time to time may be described under three groups:

(*i*) Protective measures

(*ii*) Promotional measures

(*iii*) Institutional measures

2.10.1 Protective Measures

These measure protect the small-scale sector from competition by large industries.

1. Reservation of items for the exclusive production of small-scale sector and not allowed for manufacture by large industries.
2. Tax concessions to small scale sector to help them to compete with large industries.
3. Preferential purchase of items produced by small-scale sector by government departments.

2.10.2 Promotional Measures

Assistance is given to small scale sector for its promotion and measures include as follows:

1. Supply of scarce raw materials to small scale sector at reasonal prices. Special raw material depots have been set up for quick supply.

2. Setting up common testing facilities for the benefit of a group of small-scale industries located in a particular region.
3. Assistance in securing orders from government departments through State Small-scale Industries Development Corporation.
4. Preference in land allocation and power connection to small-scale units. Power is also supplied at economical rate.
5. Setting up of industrial estates. The industrial sheds are given to small-scale units on the basis of installation.
6. Technical assistance to modernize methods of production by Central Small Industries Organization.
7. Financial assistance and provision of concessional finance through commercial banks and financial institutions.

2.10.3 Institutional Measures

An institutional network has been set up for the benefit of small-scale industries.

1. Small-Scale Industry Development Organization (SSIDO) give training and other services to run the enterprize.
2. National Small Industries Corporation (NSIC) supplies machinery on hire-purchase basis, provides marketing facilities and supplies scarce raw materials.
3. District Industries Centre have been set up in every district to serve as local point of development.
4. Khadi and Village Industries Commission (KVIC) encourages the production and marketing of khadi and handicraft items.
5. Special boards provide technical, financial and marketing facilities in their areas such as Central Silk Board, Coir Board, etc.
6. Funds for fixed and working capital and upgradation of technology and other financial assistance are provided by Small Industries Development Fund (SIDF) and Small Industries Development Bank of India (SIDBI).
7. Infrastructure for growth of small-scale industries in the semi-rural and rural areas is provided through Industrial Estates and Industrial Parks.

2.11 HIGHLIGHTS OF INDUSTRIAL AND LICENCING POLICY OF INDIA

2.11.1 Background

Industrial policy of India was based on the following factors:

1. In USA and other Western countries, the lopsided development of large-scale industries had led to imbalance in the distribution of wealth.
2. India at the time of independence was mostly rural and there was wide disparity of income between the rural rustic and urban populance.

3. The economy was predominantly agricultural, however, there were also industries producing some products like cement, cotton and jute textiles. These industries were located in small pockets owned by family groups.
4. There was a metamorphosis of economy in India from agricultural to industrial.
5. There was danger of small-scale, cottage and village sector going into oblivision and the workforce migrating from the villages to the metropolitan cities.
6. The technology appropriate for India would be simple, small and capital saving and formulated the concept of Intermediate Technology.
7. India is long in labour and short in capital. India required a level of technology, capital investment different from that current in the Western countries which were long in capital and short in labour.

2.11.2 Classification of Industries

Based on the above background, the industries to be developed in India were classified into the following categories.

1. Public Sector Undertaking

These are large scale key industries which required a huge capital outlay, high technology, advanced technical know-how and long gestation period. These industries are owned by the Government. These industries are reserved for the public sector and are listed as Annexure-II.

2. Large Industries

These are large industries owned by big business houses who had acquired experience and had capital to invest, called the MRTP companies, were allowed to invest in Hi-tech industries referred to as Annexure-III. Foreign investment was welcome in these industries freely. Industrial licencing was compulsory for these industries.

3. Small-Scale and Village and Cottage Industries

These industries have limited investment and are encouraged along with medium and large industries. These industries are planned to achieve local self-sufficiency. The small-scale sector can contribute to the growth of local economy by creating access to local markets. There are a large number of industries engaged in manufacturing, processing, preservation and servicing of goods. There are industries exclusively reserved for small-scale sector.

2.11.3. Industrial Policy Statements After Independence

Small and tiny industries have special characteristics of producing goods in a short gestation period, creating large employment opportunities at relatively low capital cost, dispersal of industrial activity and expanding the base of indigeneous entrepreneurship. Ancillary industries were included in the list of small scale industries to reduce reliance on imported components and materials required by large industries. This also led to growth of entrepreneurship. Successive industrial policies and programmes reviewed the promotion and growth of small and tiny units in the overall development of the economy.

1. Industrial Policy Statement, 1948

(*i*) The first policy statement accepted the importance of small-scale industries in the overall industrial development of the country.

(*ii*) Small-scale industries are particularly suited for the utilization of local resources and for creation of employment opportunities.

(*iii*) Small-scale industries face acute problems of raw materials, capital, skilled labour, marketing, etc.

(*iv*) The above problems of small entrepreneurs should be solved by the Central Government with the co-operation of State Governments.

(*v*) The main thrust of the policy was "protection" of small scale enterprizes.

2. Industrial Policy Statement, 1956

(*i*) The decentralized centre should acquire sufficient vitality to self-supporting. Its development should be integrated with that of large-scale industry.

(*ii*) 128 items were reserved for exclusive production in small-scale sector.

(*iii*) Small Scale Industries Board (SSIB) was constituted to examine and formulate a development plan for small scale industries during Third Five Year Plan (1961-66).

(*iv*) "Rural Industries Projects" and "Industrial Estate Projects" were started to strengthen small-scale sector.

(*v*) The main thrust of the policy statement was "Protection plus Development" of small-scale sector.

3. Industrial Policy Statement, 1977

(*i*) The small-scale sector was classified into three categories:

— Cottage and household industries.

— Tiny sector with investment in plant and machinery up to ₹ 1 lakh.

— Small scale industries with investment up to ₹ 10 lakhs and ₹ 15 lakhs for ancillary units.

(*ii*) Reservation of 504 items for exclusive production in small-scale sector.

(*iii*) District Industry Centre (DIC) in every district to promote and support requirement under a single roof.

(*iv*) The small scale sector was to be "protected, developed and promoted".

4. Industrial Policy Statement, 1980

(*i*) Increase in investment ceilings as follows:

Tiny units	– ₹ 2 lakh
Small-scale units	– ₹ 20 lakh
Ancillary units	– ₹ 25 lakh

(*ii*) Nucleus plants to replace District Industry Centres in each industrially backward district to promote maximum number of small-scale industries.

(*iii*) Promotion of village and rural industries to generate economic viability in the villages compatible with environment.

5. Industrial Policy Statement, 1990

(*i*) Increase in investment ceiling in plant and machinery

Small-scale units	–	₹ 60 lakh
Ancillary units	–	₹ 75 lakh
Tiny units	–	₹ 5 lakh

(*ii*) 836 items were reserved for exclusive manufacture in small-scale sector.

(*iii*) Scheme of Central Investment Subsidy for small-scale sector in rural and backward areas.

(*iv*) Technology Development Centre in Small Industries Development Organization (SIDO) for technology upgradation in small-scale sector.

(*v*) Small Industries Development Bank of India (SIDBI) to ensure adequate and timely flow of credit facilities.

(*vi*) Entrepreneurship Development Programmes for women and young entrepreneurs.

(*vii*) Implementation of delicencing of all units with investment upto ₹ 25 crores in fixed assets in non-backward areas and ₹ 75 crores in centrally notified backward areas.

2.12 NEW SMALL ENTERPRIZE POLICY, 1991 AFTER LIBERALIZATION OF ECONOMY

Policy measures for Promoting and Strengthening and Supplementing Small, Tiny and Village Enterprizes to deregulate and debureaucratize the small-scale sector has the following salient features:

1. Increase in the investment limit in plant and machinery of tiny enterprizes to ₹ 5 lakh irrespective of location of the unit.
2. Inclusion of industry-related services and business units as small-scale industries.
3. Limited Partnership Act to limit financial liability of new entrepreneur to the capital invested.
4. Scheme of Integrated Infrastructure Development for small-scale industries.
5. Factoring services to solve the problem of delayed payments to small sector.
6. Specialized marketing agencies for market promotion of small-scale industries.
7. Technology Development Cell in Small Industries Development Organization (SIDO).

8. Allocation of indigenous raw materials on priority to small and tiny sector.
9. Export Development Centre in SIDO.
10. Widening of scope of National Equity Fund (NEF) enlargement of single window scheme and associating commercial banks to provide composite loans to small-scale industries.

2.12.1 Comprehensive Policy Package, 2000

The following are the salient features of new policy:

1. Exemption limit for excise duty raised from ₹ 50 lakh to ₹ 1 crore.
2. Credit linked capital subsidy of 12% on loans for technology upgradation.
3. Investment limit of industry related service and business enterprizes from ₹ 5 lakh to ₹ 10 lakh.
4. ₹ 75,000 to each small-scale enterprizes to obtain ISO 9000 certification.
5. Small Scale Industries Association to develop testing laboratories with 50% reimbursement.
6. Limit for composite loans from banks raised to ₹ 25 lakh.
7. No need of collateral securities for loans upto ₹ 5 lakh.

2.12.2 Micro, Small and Medium Enterprizes Development Act, 2006

1. Investment limit raised to ₹ 5 crore.
2. SSI Excise Exemption schemes raised from ₹ 3 crore to ₹ 4 crore.
3. Credit Guarantee Fund (scheme) to provide guarantee for loans upto ₹ 25 lakh.
4. 15% capital subsidy for loans upto ₹ 1 crore for technology upgradation.
5. The Integrated Infrastructure Development (IID) schemes extended to entire country with 50% reservation for rural areas.
6. Market Development Assistance (MDA) exclusively for small-scale sector.
7. The composite loans limit has been raised from ₹ 50 lakh to ₹ 1 crore.
8. Deservation of items for small-scale units has been reduced to 35 only in 2008.

2.13 SPECIAL INCENTIVES FOR SMALL-SCALE SECTOR

2.13.1 Special Incentives and Subsidy are Given to Small-scale Sector to Achieve the Following

1. As a motivational force to attract prospective entrepreneurs to enter manufacturing or service line.
2. As an encouragement to entrepreneurs to start industries and business in backward areas.

3. Ensuring industrial development uniformly in all regions.
4. Development of new entrepreneurs to create entrepreneurial society.
5. Increasing the capability of entrepreneurs to face competition successfully.
6. Reducing problems of small scale entrepreneurs.

2.13.2 Different Incentive Schemes by Central and State Governments

Various benefits are announced from time to time and may differ from one state to another. These incentives are passed on to entrepreneurs through state level agencies.

1. Export/Import subsidies and bounties.
2. Interest free loans.
3. Subsidy for R&D work.
4. Capital investment subsidy.
5. Transport subsidy
6. Interest subsidy.
7. Subsidy for power generation
8. Exemption from property tax.
9. Subsidies to artisans and handlooms.
10. Incentives to non-resident Indians.
11. Special incentives to women entrepreneurs.
12. Exemption from Income Tax.
13. Interest fee sales tax loan.
14. Sales tax exemptions.
15. Subsidy for buying test equipment.
16. Subsidy for industrial housing.
17. Land and building at concessional rates.
18. Price preference to SSI units.
19. Subsidy/Assistance for technical consultancy.
20. Exemption from stamp duty.
21. Concessional water.
22. Provision for seed capital.
23. Allotment of controlled or subsidised raw materials.
24. Allotment of developed/constructed sheds.
25. Subsidy for cost of market study/feasibility study or reports.

2.14 THE ECONOMIC PACKAGE FOR THE MSMEs ANNOUNCED ON 13th MAY, 2020

- India's profile is based on the five Pillars of economy, infrastructure, technology, driven systems, demography and demand.
- Critical areas of economy includes land, labour, liquidity and law.
- The Economic Package announced relate to the viability and operations of MSMEs.

(*i*) Guarantee free and collateral free loans amounting to total ₹ 3.00 lakh crores. There will be a moratorium of 12 months on payment of Principal. It is expected that about 45 lakh units can take benefit, resume business activity and safeguard the employment of people.

(*ii*) As per the recommendations of the Reserve Bank of India and U.K. Sinha Committee, it is recommended to set up a Sub-ordinate Debt Fund, with a provision of ₹ 20,000 crores.

(*iii*) Fund of Funds has been setup with ₹ 10,000 crore fund. It can be leveraged for infusion of ₹ 50,000 crores through the market in the model of "Mother and Daughter Funds".

(*iv*) The definition of MSMEs upwardly has been changed as given in para 2.2.1. This will help the MSME Sector to grow as, low threshold was being a dampner for an urge to grow.

(*v*) In Government procurement, there would be no global tendering upto ₹ 200 crores. This will, help the domestic companies in general and on the other hand it will also help the MSMEs of the country in particular. These units are not able to withstand the pressure of undue competition from companies of other countries.

(*vi*) In addition to physical trade fairs and exhibitors there will be e-marketing linkage for MSME sector.

(*vii*) It is mandatory for the Departments of Government of India and the CPSEs to pay the receivables to MSMEs within next 45 days to help easing the situation of working capital of the MSMEs.

(*viii*) TDS and TCS will be reduced by 25% and all refunds to entities including Proprietorship, Partnership firms and LLP etc., woul/d be issued immediately, to help the liquidity situation of MSMEs.

2.15 STAGES IN STARTING A SMALL-SCALE INDUSTRY

There are certain procedural requirements for setting up a small-scale industry. Setting up an industrial enterprize consists of the fulfillment of a programme with the main constraints of technical, financial, time schedule and legal obligation. Every industrial enterprize is expected to ensure compliance with various enactments, regulations, legislations, etc., of central and state governments and those of local bodies before, in the course of its establishment, and after it is established.

Basic stages to start a small-scale unit can be as follows:

1. **Preliminary stage** to search a business opportunity.
2. **Decision-making stage** to select the final product, prepare feasibility project report and make all major decision regarding plant location, etc.
3. **Planning stage** for making various applications and preparing detailed project report.
4. **Implementing stage** to purchase and arrange all resources, assembly the factory and carry out production trials.
5. **Managing stage** for production and planning, inventory control, quality control, financial and marketing management.

2.15.1 Time Schedule Plan for Starting a Business

A simple and easy to use time schedule is prepared in the form of a bar chart as shown in Fig. 2.1. It should contain all the tasks to be accomplished, their order and time of starting and completion. This is a very effective management tool to complete the project successfully.

TABLE 2.2. Time Schedule Plan

Activity	Time (months) 1–2–3–4–5–6–7–8–9–10–11–12–13–14–15–16–17–18
1. Decide to go into business	
2. Analysis of strengths and weaknesses	
3. Selection of product or services	
4. Conduct market research	
5. Assessment of polential market share	
6. Site selection	
7. Preparation of financial plan	
8. Preparation of production plan	
9. Preparation of market plan	
10. Borrow funds to begin	

2.15.2 Agencies to be Contacted for Project Implementation

The agencies to be contacted for starting and implementation of the project are tabulated in Table 2.3. Single point contacts at District Industries Centre (DIC) are good place to start with the project implementation.

TABLE 2.3. Agencies for Project Implementation

Requirement	Agency/Person	Remark
Constitution of Firm		
1. Proprietorship	Legal advisor	No legal obligation
2. Partnership	Legal advisor Registrar of Firms	Agreement to be registered with Registrar of Firm
3. Company (Pvt. Ltd.)	CA/Company secretary Registrar of Company	Incorporation of company with the Registrar of Companies in the state
Registration		
1. For SSI, ancillary or cottage industry	DIC/Deptt. of Industries District Khadi and Village Industries	Temporary registration Permanent registration after establishment
2. Subsidy	DIC/IC office	Registration for subsidy for state level and central level

Contd...

3. Sales Tax (State and Central)	Sales Tax Deptt. Legal advisor	Registration for Sales Tax No. for state and interstate transactions
4. Production licence	DGTD	For units other than SSI
5. Excise	Deptt. of Excise and Customs	Registration for excise for scheduled industries
6. Factory Act	Factory Inspector Legal Advisor	Factory Act Registration
7. Shop and Establis-hment Act.	Office of Shop and Establishment	For non-manufacturing activities
8. Patents and Trade marks	Trade Mark Registry	For exclusive rights for use of mark or designs
No Objection Certificate (NOC)		
1. Location and construction	Local authorities (Municipality, Gram Panchayat, Collector)	NOC for construction
	Industrial Development Corporation	NOC for manufacturing
2. Water or Air Pollution	Water and Air Pollution Control Board	NOC
Licences & Permissions		
1. Production Programme	DI, SISI, DIC, Drugs and Cosmetics	Specializaed industries like electronics approval of Development Commission is required before establ-ishment
2. Raw material and machinery	DI, DIC, SIDC	Registration for quota after SSI registration
3. Import licence	STC/MMTC/Controller of Imports	—
4. Building plan and construction	DIC/LA	—
Special Permissions		
(*a*) Storage	Fire Brigade, Police Deptt. DC	For explosives and highly inflammable materials
(*b*) Maintenance	Chief Boiler Inspector DI/Inspector of Weight and Measures)	Boiler operation All weights to be stamped and verified before use
(*c*) Regulation of raw materials, prices distribution	Legal advisor	Flour Mills, Salt, Wood, Food, Fertilizer, Oil, Rubber, Rice Mills Insecti-cides, Jute, Mining, etc. require special permission and licensing.

Contd...

Periodic Inspection		
(*a*) Electricity rules	Electrical Inspector	Periodic inspection during production by Electrical Inspector
(*b*) Pollution Control Board	Water Pollution Treatment	Periodic Inspection
(*c*) Factory Act	Factory Inspector	Periodic Inspection
(*d*) Boiler Act	Boiler Inspector	Periodic Inspection

Question Bank

1. What is small scale industry ? Discuss the importance of small-scale industry in the economic development of a country like India.
2. What are the government policies for the development of small scale industries? Discuss.
3. How will you define a 'Small scale Industry'? Outline in brief the characteristics and types of small scale industries operating in India.
4. Discuss the various measures that have been taken by the government for promoting small-scale industrial units in India. Are these measures adequate?
5. Discuss the procedures for setting up a small scale industry. What legal obligations are to be met ?
6. Prepare a typical time schedule for starting a business.
7. What is the role of small-scale sector ? How are these industries classified?
8. What are different incentive available for small scale industries?
9. Discuss the main merits of small-scale industries. Give reasons for sickness in small industries.
10. Discuss the various policies of government towards small scale sector after independence.
11. Discuss the relation between small scale sector and large industries.
12. Discuss the growth of small scale industry in a developing country.

❑❑❑

PART – II

Business Planning

Chapter

3 Project Identification and Selection

3.1 INTRODUCTION

Entrepreneurial activity is essentially person-oriented. The make-up of the entrepreneur is based on his strengths, weaknesses, preferences, values, etc. and it will affect his choice of business opportunities. An entrepreneur therefore, should always be aware of his ownself.

Business opportunity and entrepreneurial response to it together form the foundation for the super structure of a new venture. Selection of right business opportunity is most important for a sound enterprise.

Selecting a business opportunity is one time decision and all subsequent efforts and plans will be around the product selected. A new entrepreneur will have many doubts and questions before he reaches the final decision of selecting a product.

3.2 PROJECT IDENTIFICATION

Project identification is concerned with the collection, compilation and analysis of economic data for the eventual purpose of locating possible opportunities for the investment and with the development of the characteristics of such opportunities.

An entrepreneur is an opportunity seeker. He should identify, explore and select the right opportunity. Opportunity is an attractive project idea which an entrepreneur accepts as a basis for his investment decision. A "possibility" is different from "business opportunity". Good business ideas must be capable of being converted into feasibility reports.

A good business opportunity must have two major ingredients.

1. Good market scope, i.e., a gap between present and likely demand and supply.
2. An attractive acceptable return on investment.

The business opportunity should be viable technically, manufacturing, commercially and managerially. These criteria are interlinked.

3.2.1 The Process of Selection

Selection of right business opportunity demands the following:

1. Understanding one's own capabilities, strengths, limitations and preferences.
2. Exploring all possible and suitable opportunities available within existing conditions and environment.
3. Comparative analysis of opportunities available, consistent with entrepreneurial capabilities and their relative benefits.
4. The business opportunity may be for manufacturing a product or for providing services. Start with one time of activity only.
5. The ideas about project may be generated by
   (*i*) New products/services not existing in the town, region or country.
   (*ii*) Developing existing products/services available in the market by changing design, specifications, contents, etc.
   (*iii*) Selection of products/services already manufactured.
6. It is temping to go for first group of ideas. These involve high risk taking. New entrepreneurs with little business experience can start with 3rd group.

3.2.2 Exploring Opportunities

The process of identifying opportunities requires intensive efforts and specialized skills. However, following guidelines may help to identify and access opportunities.

1. Environment

- Basic features of an area and its resources inventory.
- Population, its composition, occupational pattern, socio-economic background, etc.

2. Current Business Scene

- Present pattern of trading and business activities in the area.
- Inter-regional flow of goods, local consumption and need for products and services.
- Emerging trends and pattern of trading and business activities.
- New demand for consumption of goods and services in the area.

3. Technology Charge

- Anticipating new opportunities arising due to technology change like computers, etc.

3.2.3 Sources for Identifying Business Opportunities

1. Resource Based Ideas

- Industries based on mineral, agriculture, marine and first resources.

- Waste based products such as agro-waste, wood-waste, metal-waste etc.

2. Linkage Based Ideas

- Industries arising out of various types of linkages such as backward and forward integration from existing lines of manufacture.
- Ancillary development project
- Industries based on substitution, i.e., components obtained from outside the region or from abroad.

3. Export/Import Related Ideas

- Import substitution
- Products for export.

4. Market Shift or Growth Related Ideas

- Consumer and industrial product increase in demand due to increase in population or changes in the composition of population, purchasing power, changes in life style, etc.

5. Special Product Ideas

- Research and invention based products
- Skill/knowledge based products
- Products purchased by Institutions, Government, Hospitals, Schools etc.
- Foreign collaboration.

6. Service Sector

- Household repair and maintenance.
- Service facilities, workshops to meet the demand from industries and households.

7. Government Policies, Priorities and Plans

By exploring these sources, the entrepreneur should prepare an inventory of various project ideas.

It will be possible to identify business opportunity on the basis of available data and information. The initial process involves collection and coordination of relevant information from the right sources.

3.2.4 Criteria for Selecting a Particular Project

After getting project ideas from different sources, project profiles can be prepared by the consultants. The project profiles are used for assessment of viability of an idea. The following criteria should be considered for selecting a particular project.

1. Investment Size

The investment size will decide the promoter's contribution, term loan and licencing requirement and scale of industry, i.e., small-scale medium scale or large scale with attendant complexities.

2. Location

There are various incentives available for locating the plant in rural and backward areas. However, the location near state capital helps to attract competent managers, easy liaison with Electricity Board, State Industrial Development Corporation, Financial Institutions and other agencies.

3. Technology

The project may need, proven indigenously available technology or high technology necessitating foreign technical collaboration.

4. Equipment

The equipment selected should be of high quality and productivity and there should be no compromise in this respect. There are always serious difficulties in running a plant using poor quality equipment.

5. Marketing

The product selected should avoid cut-throat competition and should be sold to a limited number of industrial customers.

3.2.5 Importance of Project Identification

Project identification is of great importance due to following reasons:

1. Projects are the catalytic agents of economic development
2. Projects result in employment and income generation.
3. Projects initiate development of infrastructure and environment.
4. Projects accurate the process of socio-culture development.
5. Projects attract substantial financial investments.
6. Projects have beneficial consequences in the long run.

3.2.6 Constraints in Implementation of Projects

There are various internal and external constraints for the entrepreneur who venture into project implementation.

(A) Internal Constraints

Internal constraints can be limitations of management system, limitation of availability of inputs, resources and outputs. These constraints are listed below :

1. Improper preparation of feasibility reports affects the duration of project implementation.
2. Lack of project management systems, organization structure, network analysis, etc.
3. Unrealistic project objectives.

4. Non-availability of sufficient physical resources such as finance, personnel, inventories and facilities.
5. Non-availability of non-physical resources such as patents, secret processes, unique skills and experiences.

(B) External Constraints

The external constraints affecting the proper implementation of project may include Government policies, capital market, environmental situations, location of the project.

1. The project may not conform to the socio-economic objectives of the country. These factors may include nature, size and location of the project.
2. NOCs, approvals, licences, foreign collaboration, foreign exchange permit, and other government policies may hinder the implementation of the project.
3. The procedures and documents of financial institutions and banks may delay the implementation of the project.

3.3 PROJECT FORMULATION

Project formulation is the systematic development of a project idea for the final decision of investment. It is needed to safeguard against risks and difficulties in the implementation of the project. It involves step-by-step investigation and development of project idea. A team of following expertise is formed to investigate the project idea.

1. Industrial economist
2. Market analysts
3. Technologist or engineer with appropriate industrial specialization
4. Management accounting expert
5. Construction engineer.
6. Industrial engineer.

A well formulated project is the best passport for getting financial assistance in a situation of resource constraint. It will also help to get various Government clearances and sanctions.

3.3.1 Elements of Project Formulation

Project formulation is an analytical management aid to take most effective project decisions. Project formulation exercise includes the following as illustrated by a flow chart in Fig. 3.1.

1. Feasibility analysis
2. Techno-economic analysis
3. Project design and network analysis
4. Input analysis.

5. Financial analysis
6. Socio-cost benefit analysis
7. Project appraisal.

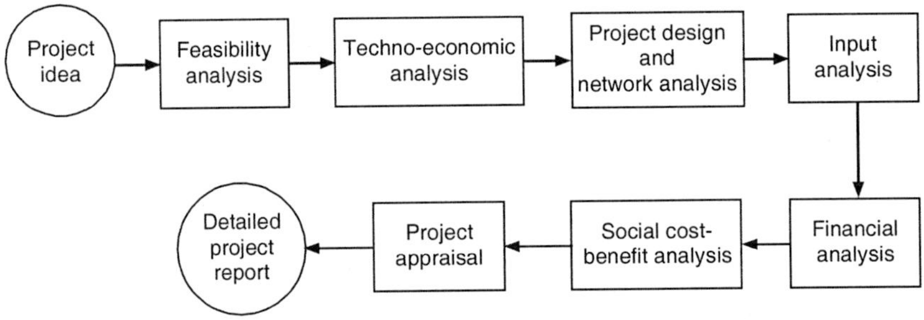

Fig. 3.1. Project Formulation

3.3.2 Feasibility Analysis

Feasibility analysis is the process of evaluating the future of a project idea within the limitations of the project implementation body and the constraints imposed on the project situation by the environment. Projects identified are normally analysed in order to establish their viability from different angles such as technical, marketing, financial, etc. The various alternatives in marketing, technology and other considerations are studied and the results alongwith supporting data are presented in a systematic form.

The project feasibility analysis is carried out in three steps :

1. Pre-feasibility study
2. Feasibility study
3. Detailed project report

1. Pre-feasibility Study

This study broadly examines the economic alternatives of the following:

(*a*) Market and plant capacity.
(*b*) Material input.
(*c*) Location and site.
(*d*) Project engineering: Technologies and equipment.
(*e*) Overheads : Factory and sales.
(*f*) Manpower
(*g*) Project implementation
(*h*) Financial analysis : Investment costs, project financing, production costs and commercial profitability.

2. Feasibility Study

This study is the detailed investigation of the project from six different angles.

(*a*) Economic
(*b*) Technical
(*c*) Managerial
(*d*) Organizational
(*e*) Commercial
(*f*) Financial

Feasibility study is an interactive process covering all aspects of an investment in project such as possible alternative solutions for production programme, location, technology, organizational set up, etc. The study should describe the optimization process, justify the assumptions made and the solution selected and define the scope of project. The viability of the project should be stated and justified.

3. Detailed Project Report

A detailed project report is prepared on the finally selected project. It is meant for entrepreneur himself and is an important document to process assistance from financial institutions and to fulfill other formalities for the implementation of the project. It is a written document and contains the following :

(*a*) Assumptions/estimates for different manpower and material needs.
(*b*) Projections : Financial, production, sales, etc.
(*c*) Information on technology, competition, prices, etc.
(*d*) Calculations : Costs, profitability, cash flow, etc.
(*e*) Documents : Quotations, land title, lease deed, partnership deed.

3.3.3 Techno-economic Study

This study is carried out for the following :

(*a*) Determination of project demand potential
(*b*) Selection of optimal project strategy.

The project design is based on the information of techno-economic study.

1. Determination of Project Demand Potential

The size of the project and the technology to be used depend on the demand potential. Demand forecasting is carried out for the following :

(*a*) It helps to firm up the quantitative parameters of the project
(*b*) It provides basis for selecting the optimal strategy for the project
(*c*) Demand forecasting may be carried out for a product or industry
(*d*) It involve determination of market characteristics, quantitative market analysis and appraisal of project demand potential.
(*e*) The data collected from various sources are compiled, tested and tabulated for interpretation and projection of future demand.

2. Selection of Optimal Project Strategy

The project formulation team determines the project size, capital investment, technology, etc. on the basis of project objectives, project feasibility parameters and project demand potential. Optimal project strategy is the combination of controlled variables to ensure achievement of project objectives with minimum expenditure of resources.

3.3.4 Project Design and Network Analysis

It consists of development of detailed work plan of the project and its time profile. This plan is presented in the form of a network diagram.

Network analysis is carried out to identify the optimal course of action to execute the project within minimum time and with the available resources. The project inputs are worked out from project design and network analysis. The financial requirements are worked out and cost-benefit profile of the project is prepared.

3.3.5 Input Analysis

The resource requirements of the project can be determined after feasibility analysis, techno-economic analysis and network analysis. Input analysis is carried out for identification, quantification and evaluation of project inputs.

The following information is obtained:

(*a*) Nature of resources to be consumed by the project

(*b*) Estimation of magnitude of required resources

(*c*) Evaluation to ensure the uninterrupted supply of inputs

Input requirements are the basis of cost estimates of the project which are used to develop financial profile of the project and carry out cost-benefit analysis.

3.3.6 Financial Analysis

Financial analysis requires estimates of project cost, services and funds requirements. It consists of break-even analysis and Ratio Analysis. The information obtained is used for commercial profitability analysis.

3.3.7 Cost Benefit Analysis

There is no standard methodology available for evaluation of social cost benefit analysis. However, it is recognised internationally. All social costs due to impact of the project are estimated alongwith secondary benefits. The analysis is carried out in same way as financial analysis. However, in place of market prices, some shadow prices are used based on social and national objectives.

The results of feasibility analysis, techno-economic analysis, design and network analysis, input analysis, financial analysis and cost-benefit analysis are consolidated and final shape is given to the project. It is also called pre-investment analysis to take investment decision about the project. This information is collected in the form of a feasibility report and is called project formulation.

3.3.8 Precautions

1. Donot select a project because someone else is doing well or someone has assumed to buy the product. Some one's assurance may be good reason to start, but not good enough to servive.
2. It is important to generate product ideas. However, many such ideas may be found unsuitable in future. It may not be possible to develop saleable product.
3. Consult experts and other people. However, all decisions must be taken by the entrepreneur.
4. It is advisable to start with a common product instead a very specialized product.
5. Check if more such projects are coming up.
6. The project should ensure future and long terms possibility of success.
7. There should be flexibility for diversion in the selected project.

3.4 FINANCING OF ENTERPRIZE

Financing an enterprize is a critical element for the success of a business or industry. Finance facilitates an entrepreneur to bring together the factors of production and produce the desired level of goods or services. The entrepreneur has to decide the need and sources of finance as per the projections in the project report.

3.4.1 Need for Finance

The amount of finance required depends upon the following factors :

1. Adequate money to purchase the fixed assets, i.e., land, building, plant and machinery, furniture, tools and gauges, etc.
2. Sufficient capital to support the operation of the business for initial three months, i.e., working capital invested in inventions of raw materials, work-in-process, finished goods, salary of the employees, payment to be made for other resources such as power, water communication, transport, etc.
3. Margin for unplanned expenses called contingencies which may constitute 10 to 15 per cent of fixed assets.

The finance needs can be classified as follows :

1. Fixed Capital or Long-term Capital

This is the money invested in some fixed assets which are required for long period of time for permanent use.

2. Working Capital or Short-term Capital

This is the money invested in current assets and is required for short period to meet day-to-day expenses.

3.4.2 Sources of Finance

The sources of finance can be broadly classified as follows :

1. Internal Sources

The funds are raised from within the enterprize, and may include :

(*i*) Owner's capital called equity.

(*ii*) Deposits and loans given by owners, partners, directors, etc.

(*iii*) Personal loans of entrepreneur from Provident Fund, Life Insurance, investment, mortgage of buildings, etc.

(*iv*) Retention of profits made and conversion of some assets into funds.

2. External Sources

The funds are raised from external sources and is called debts. This may include :

(*i*) Deposits or borrowing from relations, friends and others.

(*ii*) Borrowings from commercial banks for working capital.

(*iii*) Credit facilities from commercial banks.

(*iv*) Term-loans from financial institutions.

(*v*) Hire purchase or leasing facility from government institution.

(*vi*) Seed/margin many from financial institutions.

(*vii*) Subsidies from the government.

(*viii*) Venture capital of such institutions.

3.4.3 Capital Structure

The funds raised from internal sources are the ownership capital and is called 'equity'. The funds raised from external sources are borrowed capital and called 'debt'. Capital structure is the ratio between debt and equity capital and is expressed as debt-equity ratio. The optimum capital structure is the financing mix incurring the least cost but yielding the maximum returns.

The capital structure should have the following features :

1. Involve minimum cost and ensure maximum yield
2. Flexible to fulfil future requirements of funds.
3. Debts should be within repaying capacity of the enterprize.
4. Should ensure proper control over the affairs of the enterprize.

The capital structure depends upon the following factors :

1. **Nature of business:** such as competition, stability of markets.
2. **Size of the enterprize:** Small units need less borrowed capital whereas large units can raise funds from borrowings.

3. **Trading or equity or leverage effect :** There is more dependence on borrowed capital if rate of return on capital employed is more than rate of interest on borrowed capital.
4. **Cash flows :** There can be more borrowed capital if there is more availability of cash flows to discharge fixed obligation.
5. **Purpose of financing :** There can be more borrowed capital to buy a production asset like new machinery. Equity or owners capital should be used for unproductive expenses like spending on employees welfare.
6. **Provision for future** to change capital structure in the future.

3.4.4 Raising of Loans

Based on period, the loans are broadly classified as

1. Long-term loans
2. Short-term loans

1. Long-term Loans

There are loans taken for fairly long duration ranging from 5 years to 15 years. These are used to acquire fixed assets such as :

1. Land and site development
2. Building and civil works, internal roads and boundary.
3. Plant and machinery, spares and tools
4. Foundations and installation expenses.
5. Miscellaneous fixed assets comprising vehicles, furniture and fixtures, office equipment, etc.

The sources of term loan are :

1. Shares
2. Debentures
3. Public deposits
4. Loans from financial institution
5. Loans from commercial banks
6. Retention of profits.

2. Short-term Loans

Short-term finance is obtained for a period upto one year. These are required to meet the day-to-day business requirements.

The sources of short-term finance can be as follows :

1. Loans from commercial banks
2. Public deposits
3. Trade credit
4. Factoring
5. Discounting Bills of Exchange

6. Bank overdraft and cash credit
7. Advances from customers
8. Accrual accounts.

3.5 FIELD-STUDY AND COLLECTION OF INFORMATION

All-round commercial information is required by the entrepreneur for decision taking and preparation of project report. A detailed field study or market survey should be conducted to collect data and information.

3.5.1 Field Study

1. Field study is carried out to have advance idea of consumer acceptance of product before developing and manufacturing it.
2. It is as important as technical research and should be carried out on continuous basis to ensure continued vitality of a business.
3. It is an organized approach to reduce marketing problems.
4. Field-study is gathering, recording and analysing the utility and marketability of a product.
5. It includes collection of information on nature of demand, nature of competition, methods of marketing and other aspects of distribution of products from production to consumer.
6. It provides an effective sales forecast.
7. It promotes soundness of marketing decisions.
8. It is designed to gather data and carry out analysis to discover the market share of the product and location and types of consumers.
9. It is also the sales analysis of own sales records of markets of customers.
10. The sources of information for field-study can be :
    (*i*) Published literature
    (*ii*) Government sources
    (*iii*) Consultants.

3.5.2 Data Collection

The following data and information is collected during field study :

1. Raw Materials

(*i*) Major manufactures and suppliers of materials.

(*ii*) Lead time required to get the materials after ordering. The terms and conditions of the supply including price, tax structure, packing, payment term, etc.

(*iii*) Minimum order quantity or standard packing.

(*iv*) Free availability of raw material or any restrictions.

(*v*) Shortages created at any particular time of the year.

(*vi*) Trend of price fluctuations and availability in the last two years.

(*vii*) Government policies and expected decisions which may affect the price and availability of the materials.

(*viii*) The advantages of local purchases or purchases from abroad.

2. Machinery and Equipment

(*i*) Manufacturers and suppliers

(*ii*) Capacity specifications and brands available in the market.

(*iii*) Price, applicable taxes, transport, accessories, etc. of the machinery.

(*iv*) Requirements of motors, starters, switches, control equipment.

(*v*) Gaurantee/warranty of performance given.

(*vi*) Repair and maintenance cost annually. List of spare parts most frequently required.

(*vii*) Terms and conditions for ordering including payment, advance, delivery time, etc.

(*viii*) Inspection requirements at the manufacturer's works and after receipt of equipment.

(*ix*) Market standing of manufacturer alongwith list of parties whom same machines have been supplied.

(*x*) Special transport and handling requirements.

(*xi*) Overall dimensions of machines and foundations and space requirements.

(*xii*) The output capacity and average (maximum) quantity of the goods produced on the machines.

3. Market Demand of Product

(*a*) Competitor Manufacturers

(*i*) General factory layout and space occupied by similar enterprise.

(*ii*) Range of products, installed capacity, utilized capacity and selling price of the competitor.

(*iii*) Normal terms and conditions of business.

(*iv*) Future plans for expansion and diversification of the competitors.

(*v*) Salient features like technical skills, finance and other resources.

(*vi*) Market share, region and practice.

(*vii*) Special problems faced by competing units.

(*b*) Trader or Supplier

(*i*) Present major traders of the product, range of products, general business terms, commission etc. of the traders.

(*ii*) Market area and annual turnover.

(*iii*) Duration and type of relation between supplier and manufacturer.

(*iv*) Possibility of trading with him and his expected business terms.

(*v*) Normal level of stock he is maintaining and any specific problems faced by him.

(*vi*) His future predictions/comments on business conditions

(*c*) Customers or Consumers

(*i*) Annual consumption and requirements.

(*ii*) Present sources of supply.

(*iii*) Customers loyalty of brand (make) and preferences like price, quality, payment term, etc.

(*iv*) Level of satisfaction with the present product and supply.

(*v*) His purchasing criteria and purchasing power.

(*vi*) Consumption pattern and basis for estimation of his requirement.

(*vii*) Expected changes in future consumption in quality due to changes in technology.

(*viii*) Size of his average order, time and frequency of the order.

3.6 THEORIES OF INDUSTRIAL LOCATION

3.6.1. Factor Affecting the Site Selection of Plant

The location of an enterprize has to be carefully selected. A wrong choice sometimes proves fatal for the enterprize. The selection of a suitable locality depends on a number of factors. Further, each factor is influenced by a number of other factors. So a thorough analysis of the location to be selected is to be undertaken before a particular site is finalized. Following list has been prepared by the United States Department of Commerce to help the site selection.

I. Location of Production (Raw) Materials

A. What production materials exist in the area?

B. In what volume are the production materials available?

C. What is the quality of each production material?

D. How accessible are the available material?

Does the nature of such production materials (quantity, bulk, perishability, etc.) or any unusual advantage attendant thereto definitely favour plant location at or near site or source?

II. Labour

A. What is labour force of the area?

B. What is the quality of the available labour force?

C. What is the character of the labour force?

D. What quantity of labour is seasonally available?

1. Is a large supply necessary?
2. Are special skills required?

III. Sites

A. How much land is available for industrial expansion?

B. What are the soil and topographic features?

C. At what cost and terms can land be purchased or leased?

D. What facilities are now or will be available?

E. What industrial floor space is or will be available in existing structures?

1. Is a large area needed?
2. Must the site have special features, such as topographic soil conditions, transportation facilities, low costs, etc.?

IV. Industrial Fuel

A. What industrial fuel is available for additional industry?

B. How dependable is the supply?

C. At what cost is the fuel available?

1. Is low cost heat a major factor in processing?
2. Is a particular fuel required?

V. Transportation Facilities

A. What transportation facilities are available?

B. What is the rate situation?

1. Do materials and/or products require prompt and/or low cost movement?
2. Does product or raw material require a special type of transportation?

VI. Markets

A. What is the trading area generally served by this location?

B. What is the general quality of the market?

1. Are time and distance vital matters to reach the market?
2. Is price of quality a vital market factor?

VII. Distribution Facilities

A. What agencies and services are available?

1. Are there channels for marketing complex and do they require special financing, inspection, warehousing, storage or terminal facilities?

VIII. Power

A. What kind of industrial power is available?

B. How dependable is the power available?

C. At what cost is the power available?

1. Is power a significant factor?
2. Is sufficient surplus power available?

IX. Water

A. How much water is available for additional industry?

B. What is the quality of the water supply?

1. Is significant quality of low cost water required in processing?
2. Whether the required quantity is available?

X. Living Conditions

A. Are adequate facilities available for education, recreation, shopping, religious and social life, health and professional services?

B. Is sufficient housing available?

1. Will the industry employ enough people to make housing and environment significant factors?

XI. Laws and Regulations

A. What law and regulation exist of significance to prospective industry?

1. Is the industry subject to special local and/or state regulations?

XII. Tax Structure

A. How favourable is the tax structure to industry?

1. Is the capital investment of such a size that the tax rate or assessment policy will be important?

XIII. Climate

A. What are the climate characteristics of the area?

Optimum site is selected on the basis of a comparative survey of the alternative sites in question.

3.6.2 Steps in Enterprize Locations

The following steps should be followed for choosing a plant location:

(*i*) Selection of the region.

(*ii*) Selection of locality.

(*iii*) Selection of the exact site.

Different surveys as per # 3.6.1 should be conducted and significant factors should be analysed. The important factors in the selection of region, locality and site are tabulated in Table 3.1.

1. Availability of Raw Materials

The region provide at least greater part of raw material required to ensure a continuity of supply at reasonable price.

TABLE 3.1. Important Factors in the Selection of a Site

S. No.	Factors	Selection of region	Selection of locality	Selection of site
1.	Availability of raw materials	✓		
2.	Nearness to markets	✓		
3.	Availability of power	✓		
4.	Transport facilities	✓		
5.	Suitability of climate	✓		
6.	Government policy	✓		
7.	Competition between states	✓		
8.	Availability of labour		✓	
9.	Civic amenities for workers		✓	
10.	Finance and research facilities		✓	
11.	Availability of water and fire-fighting facilities		✓	
12.	Local taxes and restrictions		✓	
13.	Momentum of an early start		✓	
14.	Personal factors		✓	
15.	Soil, size and topography			✓
16.	Disposal of waste			✓
17.	Housing facilities for workers			✓
18.	Unecumbered land			✓
19.	Good scenery			✓
20.	Road, rail and water connections			✓

2. Availability of Skilled and Unskilled Labour

Proper labour required should be available in the area. It is costly to bring labour from other areas.

3. Nearness to the Source of Motive Power

The area should provide cheap and reliable power, soft water and adequate sewage disposal facilities.

4. Nearness to Market

Nearness to market has to be taken into account.

5. Availability of Transport Facilities

Transport should be available at a reasonable cost.

6. Nuisance Problems

An arrangement has to be made, if not available, for smoke, noise, odour or smog existing in the area.

7. Suitability of Climate

Climate does influence the product and its quality.

8. Topography

Flat, cheap, marshy land, sea, lake, river location, vegetation in the area, elevation above sea level.

9. Soil Condition

This determines the nature of foundation, drainage requirements.

10. Water Supply

Quality of water is important. Softening and treatment plant, if required, will have additional cost.

11. Waste Disposal

Local regulations may require additional facilities for treatment of effluents prior to discharge to main stream.

12. Nearness to Rail Connection or Highway

13. Local Legislation and Municipal Laws

Laws for sanitation, air and steam pollution.

14. Tax concessions and other incentives

15. Community attitude

Towards manufacturing activities and outside investors.

16. Living conditions for labour and management

Climate, recreational facilities, quality of schools, religious institutions, medical facilities, hospitals and other factors.

"Most advantageous location is that at which the cost of gathering material and fabricating it plus the cost of distributing the finished product to the customer will be minimum". Kimbal and Kimbal.

3.6.3 Analysis of Site Selection

Four technical entrepreneurs formed or private limited company to manufacture fine chemicals. The tangible and intangible factors related to four locations studied are tabulated below. What will be your recommendation?

TABLE 3.2. Potential Plant Sites

Factors	Location			
	A	**B**	**C**	**D (Rs. in lakh)**
Tangible Factors				
1. Land and land development	1.00	1.20	1.50	0.70
2. Building	7.00	8.00	7.50	8.00
3. Plant and machinery as selected	33.00	32.00	34.00	33.00
4. Annual cost of raw materials	20.00	24.00	26.00	22.00
5. Annual salaries and wages	8.00	8.00	8.50	9.00
6. Annual utility charges	7.00	5.00	4.00	3.00
7. Distribution expenses (annual)	3.00	1.00	2.50	0.50
8. Annual sales	60.00	53.00	55.00	60.00
Intangible Factors				
1. Labour attitude	Poor	Indifferent	Good	Good
2. Recreational facilities	Good	Good	Good	Bad
3. Community facilities	Fair	Poor	Fair	Good
1. **Total expenses** [4 + 5 + 6 + 7]	38.00	38.00	41.00	34.50
2. Profit [Sales – Total Expenses]	22.00	15.00	14.00	25.50
3. Total Investment [1 + 2 + 3]	41.00	41.20	43.00	48.00
4. Return on Instrument [%] $\left[\dfrac{\text{Profit} \times 100}{\text{Total Investment}}\right]$	53.60	36.41	32.56	53.13

Locations A and D are attractive based on Return on Instrument final decision between location A and D can be decided on intangible factors.

	Weightage				
1. Attitude of labours	2	5	—	—	2
2. Recreational facilities	3	2	—	—	8
3. Community facilities	2	4	—	—	2
4. Relative merit of Site A to D	$\left(\frac{5}{2}\right)^2 \times \left(\frac{2}{8}\right)^3 \times \left(\frac{4}{2}\right)^2 = \frac{25}{4} \times \frac{8}{512} \times \frac{16}{4} = \frac{200}{512}$				

Site A is preferred to site D on intangible factors.

Question Bank

1. What is project identification ? Explain the internal and external constraints in project identification.
2. Outline and discuss briefly the various steps involved in the identification of a project.
3. Explain the various stages of project formulation.
4. "Project formulation is a screening process through which a project idea is developed systematically and effectively." Discuss this statement.
5. Explain the need of project financing. What are sources of finance ? How is capital structure decided ?
6. What is the importance of field study ? What data and information should be collected for preparation of a project report ?

❐❐❐

Chapter

4

Project Report Preparation

4.1. INTRODUCTION

Soon after the identification of a project, a project report is formulated after examining various relevant aspects such as economic, technical, financial, product and managarial aspects. The project report is submitted to financial institutions and commercial banks for getting loans. The report is analysed by these institutes to see how production should be organised for maximum returns. Project report is of great importance. It highlights the practicability of a project in terms of different factors like economy, finance, technology and social desirability. An important aspect of the report is in determining the profitability of the project and minimising risks in the execution of the project.

There are three stages of a project report.

1. Preliminary Project Report (PPR).

   A brief outline of the project justifying its viability.
2. Techno-economic feasibility assessment.
3. Detailed Project Report

4.2. PRELIMINARY PROJECT REPORT

The project report must be prepared by the entrepreneur himself to run the industry successfully. Even when he takes the help of a consultant, he must involve himself in the preparation of the Project Report.

Preliminary Project Report is simple brief data sheet which gives insight into the following aspects of the project.

1. Requirement of amount of money, manpower and materials for setting up the project.
2. Types of machines required.
3. Sources of technology.
4. Economic gains from the project.

4.2.1. Advantages of Preliminary Project Report

An entrepreneur selects 3 to 4 items (and not one) initially. It is very costly and time consuming to prepare detailed project report for all the items selected. Therefore preliminary project reports are prepared which are very short. There are other advantages also.

1. A preliminary project report has sufficient data to get provisional registration with the state government. The planning and execution can be started. At least action can be taken on activity requiring long periods.
2. Certain formalities can be completed on the basis of data in the preliminary project report. For example, production programme for an electronic unit can be approved by the state/central government because this approval requires only projected production level and raw material requirements.
3. The data collected for preparing a preliminary project report can be used to start preparing detailed project report.
4. Preliminary project report helps identify infrastructual requirements for the project such as land or shed. Accordingly applications can be made to concerned government agencies in advance.
5. The preliminary project report instills confidence and motivation in the entrepreneur to start the time consuming process of collection of data for preparation of a detailed project report.

4.2.2. Preparation of Preliminary Project Report

A sample proforma is given in Table 4.1 for the preparation of preliminary project report systematically. The information needed could be collected and presented quickly.

The methodology is discussed below. Apprehend market survey report with the preliminary project report. The report can be used for further actions.

TABLE 4.1. Preliminary Project Report Proforma

1.0. General

Name of Entrepreneur : ____

Birth Date : ______________________________ Age : ____________________

Project ____

Location ____ /Rented shed.

Type of Organisation : Proprietary/Partnership/ ______________________

Name of Firm ____

Address ____

____

____

1.1. Eductional Qualifications

S.S.C. or below	Degree/Diploma	Institute	Major Subject	Year of Passing

1.2. Special Training

Training Subject	Institute	Duration	Achievements

1.3. Work Experience (Past and Present)

Organisation	Position	Nature of work	Duration

2.0. Details of Proposed Project: Manufacturing/Servicing

2.1. Production Programme

S. No.	Items	Total Quantity Per Year	Annual sales Revenue	Capacity Utilization

2.2. Machines/Equipment

S. No.	Description	No.Regd.	Price	Total Value	Names and Address of suppliers
	Total				

2.3. Raw Materials

S. No.	Item	Total Annual Requirement		Source
		Quantity	Value (Rs.)	
	Total			

2.4. Utilities

S. No.	Description	Annual Requirements	Total annual expenses (Rs.)	Remarks
1.	Electricity			
2.	Water			
3.	Coal/Oil			
4.	Others			
	Total			

2.5. Man Power

S. No.	Description	No.	Total Wages and Salaries Per Yer (Rs.)	Remarks
1.	Skilled			
2.	Semiskilled			
3.	Unskilled			
4.	Office staff			
	Total			

3.0. Market Study: Use Separate Sheets

4.0. Cost of Project

4.1 Fixed Capital

S. No.	Description	Value (Rs.)
1.	Land/Building	
2.	Machinery/Equipment	
3.	Furniture and Fixtures	
	Total	

4.2. Working Capital

S. No.	Description	Duration	Quantity	Value (Rs.)
1.	Raw Material Stock			
2.	Semi-finished goods stock			
3.	Finished goods stock			
4.	Monthly production expenses utilities + wages + salaries			
	Total			

4.3. Total Project Cost

S. No.	Description	Value (Rs.)
1.	Fixed Capital	
2.	Working Capital	
3.	Preliminiary and pre-operative expenses	
	Total	

4.4. Means of Finance

S. No.	Description	Value (Rs.)	Remarks
1.	Term Loan		
2.	Working Capital Loan		
3.	Own investment		
4.	Subsidy		
5.	Any other		

4.5. Profitability Analysis

S. No.	Description	Value (Rs.)
1.	Sales Revenue	
2.	Manufacturing expenses (2.3 + 2.4 + 2.5)	
3.	Selling and Distribution expenses	
4.	Administrative expenses	
5.	Interest	
6.	Deprecition	
7.	Gross Profit [1- (2 + 3 + 4 + 5 + 6)]	
8.	Income Tax	
9.	Net Profit (7-8)	

5.0. Supplementary Details

5.1 Own House/Property, etc.

5.2 Own Insurance Policy.

5.3 Interest in other firms.

5.4 Tick whichever is applicable

S.C./S.T./O.B.C./General

5.5 Present monthly income.

6.0 REFERENCES

S. No	Name	Address	Occupation

Date : Signature :

Place :

4.3. TECHNO-ECONOMIC FEASIBILITY STUDY

1. Location

The location should be decided on the following basis:

- Availability of raw material
- Availability of labour
- Availability of market outlets
- Cost of transportation of raw materials and finished goods
- Place self owned/rented premises/industrial area.
- Requirement of size of land/shed.

2. Educational Qualifications and Experience

This information and background of entrepreneur will help in the acceptance of application by financial institution.

3. Production Programme

The production programme is worked out on the basis of data and information obtained from market survey and market research. The missing information may be obtained from the consultant.

Readymade project profiles are available with DIC (District Industries Centre), good libraries or consulting organisations. They can be good source of information. Visit can be made to an existing unit for collection of data.

4. Plant and Machinery

The list and specifications is based on data collected by market survey. The information collected should also include power requirements. The cost of machinery will include ex-works cost, transportation, handling, insurance, taxes, foundations, erection and commissioning charges and cost of spares. Or 20% cost may be added to the quoted price for rough estimates of the cost of plant and machinery.

5. Plant Capacity

The quantity of goods can be worked out from the capacity of the machinery, normal working days (300 days per year) and man hours lost due to breakdown and planned maintenance : 70% to 80% of equipment capacity can be taken as plant capacity.

6. Sales Revenue

The sale price should be decided on the basis of market price of the product, minus commission and discount, etc. The sales revenue will be product of annual production and sales price.

7. Raw Materials

The raw material list will include raw material for production, packing, maintenance and testing: The details can be collected by market survey. The item-wise requirement of raw materials in terms of quantity and value should be given. Similarly the sources of procurement of raw material have to be indicated.

8. Utilities

Electricity, coal, furnace oil, diesel/petrol, compressed air, water, etc. are required to facilitate production. The electricity requirements should also include power required for running motors, heating and lighting.

9. Manpower Requirements

The manpower required will be for

(*a*) operating machinery

(*b*) assembly of final product, packing and supervisions

(*c*) selling/marketing staff

(*d*) office works

The salaries and wages have to be decided as paid by other manufactures and also as per Minimum Wages Act. Add 20% to salaries to account for additional benefits for the workers.

10. Market Study

Market survey report should be attached to the preliminary project report.

11. Cost of Project and Profitability

The calculations involved are:

(*a*) The cost of purchase and installation of plant and machinery.

(*b*) Cost of land/shed purchased on ownership/lease basis.

(*c*) Technical know-how cost.

(*d*) Expenses towards preparation of project report, market survey, travelling for data collection, deposits to electricity board, telephone department, etc. These expenses together are called preliminary and pre-operating expenses.

12. Working Capital

This includes value of assets.

(*a*) Stock of raw materials

(*b*) Stock of finished goods

(*c*) Semi-finished goods or goods under process

(*d*) Receipts from customers against orders

(*e*) Cash to meet day-to-day expenses

The stock of raw materials will depend upon the annual production capacity, lead time for the procurement of raw materials.

The stock of finished goods will depends upon the requirements of buyers, economic size for transportation. The cost of stock of finished goods can be calculated on the basis of ex-factory price.

The amount of goods under process will depend upon the production cycle for converting raw materials into finished good. The cost of this stock can be taken as arrange of cost of raw materials and ex-factory price of products.

Normally credit facilities range for 30 to 50 days and it can be safely taken that payment for goods sold will be only after 45 days. Total debtors can be calculated from annual production in terms of quantity and the value for 45 days period.

13. Project Cost

The total project cost is the sum of fixed capital and working capital.

14. Finance of Sources

(*i*) 75 to 80% of fixed capital can be financed by central and sate financial corporations in the form of long-term loan.

(*ii*) 60 to 75% of working capital would come from commercial banks as working capital loan.

15. Project Profitability

(*i*) Manufacturing expenses = Cost of raw materials + Utilities + Manpower.

(*ii*) Selling and distribution expenses include commission payable to salesman + expenses towards advertisement (lumpsum) + publicity expenses.

(*iii*) Administrative expenses include expenses on postage, stationery, telephone and telegraph charges.

(*iv*) The interest has two parts: Interest on Term Loan and interest on working capital.

The interest on term loan can be 12.5% to 14.5% depending upon quantum of loan, lending agency and scheme of loan. The interest on working capital loan can be calculated @ 13.5% to 16.5%.

(*v*) A flat rate of 15% of value of machinery and 5% of value of building can be taken towards depreciation.

(*vi*) Miscellaneous expenses can be estimated on lumpsum basis.

(*vii*) Total cost will be sum of all above expenses.
(*viii*) Sales Revenue is already calculated.
(*ix*) Gross Profit = Total Sales Revenue – Total Cost.
(*x*) Tax should be calculated from current tax adviser.
(*xi*) Net profit = Gross Profit – Tax.

16. The details of the promoters are relevant to financial institutions.

4.4. DETAILED PROJECT REPORT (DPR)

A detailed project report has to be prepared on the finally selected project. It is meant for entrepreneur himself and is an important document to process assistance from financial institutions and to fulfill other formalities for the implementation of the project. A well perceived, well made project report by the entrepreneur himself is helpful to him for running the industry effectively. Therefore, even when he choses to take the help of a consultant, he must involve himself in the preparation of the project report.

4.4.1. Contents of Detailed Project Report

A detailed project report can be used for the following:

(*a*) It indicates cash flow priorities.

(*b*) It helps to realise the resources for the implementation of the project.

(*c*) It gives a general idea of resource requirement and means to procure them.

(*d*) It shows the feasibility and probability of achieving profit.

(*e*) It indicates likely returns and benefits from the unit.

A detailed project report helps to visualise financial rewards, needs, commitment and actions.

The project report is a written document and consists of the following:

1. **Assumptions/Estimates**

For production, sales, raw materials consumptions, manpower needs, prices, etc.

2. **Projections**

Financial, production, sales, etc.

3. **Information**

On technology, competition, prices, etc.

4. **Calculations**

Costs, profitability, cash flow, etc.

5. **Documents**

Machinery and raw material price quotations; constitutional information such as partnership deed, documents for infrastructural need, land building, land title, lease deed, rent deed, etc.

4.4.2. Preparation of Project Report

1. **Introduction of the Project Selected**
   (*a*) The name of product selected
   (*b*) Application of the product
   (*c*) Future demand of the product
   (*d*) Demand and supply position of the product, reasons of selection, import substitution, export product, general prospects of commercial success.

2. **Entrepreneur's Bio-data**
   (*a*) Personal information Name, address, age, family background.
   (*b*) Education qualification and training.
   (*c*) Work experience, specialised capabilities and other relevant data.

Note: If education, training and work experience are related to the product selected, the relevance of experience to the project should be explained.

3. **Market Analysis**

Market survey report covers the competitiors, machinery/raw material suppliers, traders, clients and their future requirements. The information can be given in table/chart form. Letters from future clients, machinery/raw material supplies should be included to make the project report more credible.

4. **Layout of the Plant**
   (*i*) Proposed constitution of firm, i.e., proprietorship/partnership, limited company, etc.
   (*ii*) Reasons for site selection.
   (*iii*) Details of facory shed rented/ready.
   (*iv*) Details of total area with expansion provisions.
   (*v*) Details of electrification of factory.
   (*vi*) Plant layout.
   (*vii*) Expected cost of land and building.

5. Project Engineering

5.1 Preliminary Information

- Financial requirement
- NOC from DIC
- Sales Tax registration
- Excise duty registration
- Factory Act registration
- Need for trial production and arrangements made.
- Calculation of preliminary and pre-operative expenses, expenses incurred before trial production, advertisements.
- expenses for getting loans, furniture and fixtures and other expenses.

5.2. Production Programme

Based on market estimates, production programme for first three years should be indicated including number of shifts. A chart can be attached to indicate the information.

5.3 Production Procedure

The following details should be given:

- Technical specification of the product
- Step-by-step manufacturing process
- Plant layout
- Quality control

5.4. Raw Materials

The report should include principal raw materials and their specifications.

- Availability of raw materials, its quantity and flexibility of procuring the same regularly,
- Calculation of the required quantity of raw material stock according to availability of the same,
- Total value of the raw material needed in one year; including transportation charges, octroi charges as shown in the following table:

S. No.	Description	Rate/Unit	Qty.	Total value
1.				
2.				
3.				
4.				
5.				

5.5. Consumables and Utilities

- Value of material to be consumed in manufacturing.
- Cost of utilities like power, water, fuel, coal/oil etc.

5.6. Machinery and Equipment

The cost estimates for plant and machinery should be given using same table as for raw material. The value of machinery and equipment should include sales tax, octroi, transportation and insurance cost. An escalation of 10% may be provided in the price.

- Installation expenses including cost of foundation for the machinery should be mentioned separately.
- Production capacity per hour for every machine should be mentioned.
- Quotations should be obtained from 3 reputed/recognised parties for comparative study.

5.7. Manpower Requirements

1. Factory staff, their number, monthly and yearly emoluments.
2. Administration staff, their number, monthly and yearly emoluments.
3. Sales staff, their number and yearly emoluments.

If special technical personnel are needed their functions and availability should be explained.

5.8. Other Administrature Expenses

This should include:

1. Stationery and printing
2. Postage
3. Telephone and trunk calls
4. Travelling
5. Business magazine subscriptions
6. Technical and legal fees.
7. Advertisement expenses
8. Monthly payment to salesmen
9. After-sales service
10. Miscellaneous expenses

5.9. Working Capital

The calculations should be made for the following:

1. Raw material stock (Days/week) ————
2. Goods under process (Days/week) ————
3. Finished goods stock (Days/week) ————
4. Stores and spares
5. Monthly expenses

5.10. Total cost of the project

1. Fixed Capital: Land and Building + Plant and Machinery
2. Preliminary and pre-operative expenses
3. Working capital
4. Provision for contingencies

5.11. Financing of the Project

1. Own investment
2. Capital subsidies from state/central government
3. From friends and relatives
4. Long-term loan from banks/corporations
5. Short-term loan from banks
6. Trade credit

   Total financing must match total project cost.

4.5. PROJECT VIABILITY

The project viability as given below should be part of the detailed project report.

1. **Profitability Calculations**

   Estimated profitability and financial ratios are given for three years.

S. No.	Description	1st Year	2nd Year	3rd Year
1.	Sales Revenue			
2.	Estimated Production Expenses			
3.	Production Profitability (1 – 2)			
4.	Other income			
5.	Profit before Tax (3 + 4)			
6.	Expected income tax to be paid			
7.	Net profit (5 – 6)			
8.	Net cash returns (7 + depreciations + investment allowance)			
9.	Percentage of net profit (*i*) On sales $\left\| \dfrac{(7)}{(1) \times 100} \right\|$ (*ii*) On investment			

2. **Break-even Point (Calculation and Explanation)**

$$\% \text{ of Breakeven} = \frac{\text{Fixed Expenses}}{\text{Expected Contribution}} \times 100$$

where Contribution = Income from sales – Variable Expenses.

3. **Cash Flow**

S. No.	Description	1st Year	2nd Year	3rd Year
(*a*)	**Financial Inflow**			
1.	Net profit (income tax and interest added but depreciation substrated)			
2.	Own investment			
3.	Depreciation			
4.	Investment allowance			
5.	Long-term loan			
6.	Short-term loan			
7.	Other income			
(*b*)	**Financial Outgo**			
1.	Preliminary and pre-operative expenses			
2.	Fixed investment			
3.	Increase in current assets			
4.	Reduction in long-term loan			

(Contd.)

S. No.	Description	1st Year	2nd Year	3rd Year
5.	Reduction in short-term loan			
6.	Interest			
7.	Income tax			
8.	Total			
(*c*)	**Repayment of Loans**			
1.	Net profit (income tax debited)			
2.	Entrepreneurs drawings			
3.	Available profit (1 – 2)			
4.	Amount to be returned to financial institutions			
5.	Interest on funds availed from state corporation and banks			

4. **Project Implementation Planning (Time in Weeks)**

(*i*) Preliminary survey of product.
(*ii*) Market research and preparation of project report.
(*iii*) Loan application, preparation and sanction.
(*iv*) Selection of site and other necessary formalities.
(*v*) Order for plant and machinery.
(*vi*) Construction of factory building (rented building).
(*vii*) Installation of plant and machinery.
(*viii*) Selection of personnel.
(*ix*) Application and formalities for electricity, fuel.
(*x*) Other necessary licenses and sanction letters necessary for project implementation.
(*xi*) Electrification in factory.
(*xii*) Trial production.
(*xiii*) Commercial production.
(*xiv*) Training of personnel.
(*xv*) Inauguration of factory.

Prepare bar chart, procedence diagram or PERT–CPM charts.

5. **Enclosures**

1. Education certificates
2. Experience certificates
3. Testimonials for licence and permission letters.
4. Drawings of factory building and expected cost of building or rent receipt.
5. Quotation of machinery and equipment (in triplicate)
6. Letters from raw materials suppliers (if possible)
7. Partnership deed or any other documents.

6. **Financial Features at a Glance**

S. No.	Description	Own Investment	Long-term loan	Short-term Loan from Bank	Total
1.	Land and Building				
2.	Plant and Machinery				
3.	Working Capital				
4.	Preliminary and pre-operative expenses				
5.	Total				
6.	Expected total production expenses				
7.	Profitability				
8.	Expected net profit (less income tax)				
9.	Break even point (%age)				

7. **Precautions**

(*i*) Always estimate cost on high side and income on lower side.
(*ii*) Use only personally verified costs and prices.
(*iii*) Data taken from other project report must be updated.
(*iv*) Make escalation provisions due to planning period and implementation period.
(*v*) Take 50 to 60% capacity utilization in 1st year, 80% in 2nd year.
(*vi*) Project report should be prepared personally.
(*vii*) Counter check all calculation.
(*viii*) Take advice and opinion of expects.

8. **Reasons of Failure of Project Report**

(*i*) Over estimation of capacity utilization.
(*ii*) Under-estimation of provisions for contingencies.
(*iii*) Under-estimation of cushion for fixtures and electrification
(*iv*) Under-estimation of cushion for handling and time loss
(*v*) Unrealistic report to present on rosy picture.
(*vi*) Without updated and relevant information and documents.
(*vii*) Overestimation of income and under-estimation of expenses.

4.6. PROJECT APPRAISAL

Project appraisal is an exercise by lending financial institutions for objective assessment of various aspects of a project. This is carried out to determine the viability of the project before allocating term finance. The various aspects appraised are shown in Fig. 4.1.

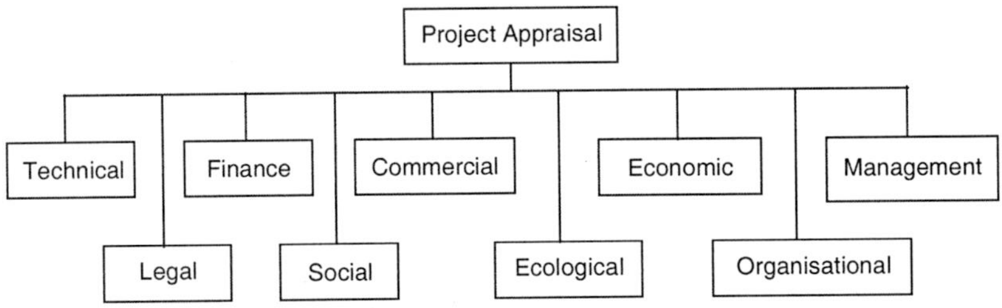

Fig. 4.1. Project Appraisal Aspects.

The scope of appraisal consists of:

- Choice of technical process and/or appropriate technology.
- Technical collaboration arrangements, if any.
- Size and scale of operations.
- Locational aspects of the project and availability of infrastructural facilities.
- Selection of plant, machinery and equipment together with background, competence and capability of suppliers.
- Plant layout and factory buildings.
- Technical engineering services.
- Project design and network analysis for the assessment of project implementation schedule.
- Effluent disposal, management of entry, utilization of by-products, etc.
- Project cost and its comparison with other similar projects based on technology, equipment, product mix and time spread.
- Project cost estimates, profitability projections, etc.
- Sensitivity analysis.

The main stages of the system of project appraisal are:

1. Economic : Indicates priority use.
2. Technical : Involves scale of project and the process adopted.
3. Organisational : Suitability is examined.
4. Managerial : Adequacy and competence are critically scruitinised.
5. Operational : Capability of the project.
6. Financial : Determines the financial viability for sound implementation and efficient operations.

Project Appraisal Format is given below:

S. No.	Criteria	Project-I	Project-II	Project-III	Project-IV
1.	Investment size				
2.	Location				
3.	Technology				
4.	Equipment				
5.	Marketing				
6.	Power and water				
7.	Other's performance				
8.	Working capital needs				
9.	Labour component				
10.	Economic viability				
	Total				

Point scale : A = 5 points, B = 4 points; C = 3 points; D = 2 points E = 1 points

4.7 DEMAND ANALYSIS

The economic and financial viability of a project depends upon scale of operation and size of the plant. The economic size of the project can be found out by analysis of capital and operating cost as function of plant capacity.

The economic plant capacity depends upon :

(*i*) Market demand

(*ii*) The technological size of standard and available equipment for specific process.

However, if the plant size is sub-optimal, the operation will not be economical and will suffer due to dis-economics of scale. The main reasons of selection of plant capacity below optimal can be.

1. **Demand Constraints :** The market demand may be too low.
2. **Resource and Input Constraints :** The available raw materials in a given region may not be sufficient to feed a large plant.
3. **Financial and Managerial Capabilities :** There can be limitation on the financial and managerial capabilities of the owner.
4. **Special Problems of fabrication** of equipment, transportation and erection of equipment may limit the size of operation.
5. **Risk of Rapid absolescence** of technology may discourage the establishment of large scale facilities.

Special precautions are needed for operating the plant below economic size.

(*i*) Capital analysis of the project

(*ii*) Performance analysis of existing plants operating below economic size.

Flexibility should be added to production facilities to change the product mix (size and quality of product) as per requirements of the market and customer. However, there will be additional cost involved in arranging flexibility in production capacity. The additional cost can be justified by carrying out cost benefit analysis on long-term basis.

Capacity planning is to integrate the factors of production to minimize plant costs over the life of a product. The monetary commitment to new production can be reduced by limiting initial investments in production machinery, holding material purchases to absolute minimum levels, and employing temporary labour. In order to design a new plant it is necessary to determine a realistic sales goal. The total capacity should be large enough to recover investments and on the other hand it should be well within market potential. Therefore, aim of capacity planning is to balance revenue and production expenses or selling price and production costs.

The following methods are used for demand analysis.

1. Material balance and output method.
2. Benefit cost analysis method
3. Discounted cash flow method.
4. Internal rate of return method.
5. Net present value method.

4.7.1 Material Balance and Output Method

A block diagram representing an industrial enterprize is shown in Fig. 4.2. The diagram shows many examples of conversion of input materials into outputs, i.e., demands are transformed into useful products, materials are converted into products with the help of machines, data is interpreted into knowledge, skills are used to convert energy into services.

It may be noted that in industrial production systems, the inputs account for variable costs of production, the conversion facilities are associated with fixed cost and outputs products are the revenue. The profit depends upon the relationship between revenue, variable and fixed costs. In other words, profit is the interaction of input and conversion costs to output revenue.

Input	Conversion system	Output
Demands	Transformation	Useful products
Materials	Machines	Products
Data	Interpretation	Knowledge
Energy	Skills	Services
Variable cost	Fixed cost	Revenue

Fig. 4.2. Block diagram of a production system

The capacity planning, i.e. The number of units made and sold should be based on the economic consideration so that average total cost of production should be minimum for the given output as shown in Fig. 4.3.

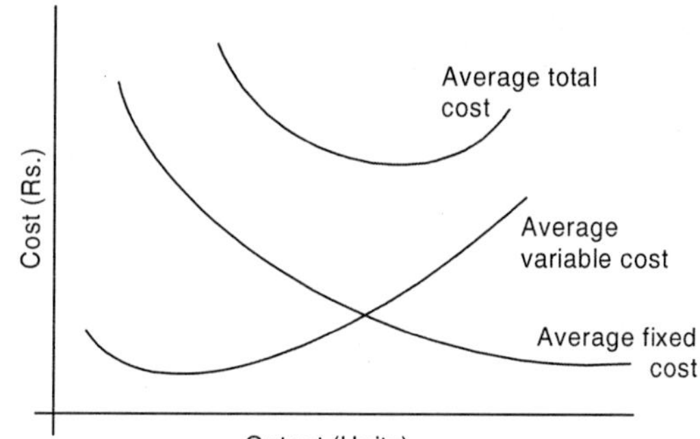

Fig. 4.3. Economic relationship

The share of average fixed cost per unit production decreases with volume of production. Business people say they spread the over head by selling more. Average variable costs are high for initial productions and decreases with increase of output due to quantity discount on materials and higher labour productivity. The average total cost is the sum of average fixed cost and average variable cost. It shows the combined effects of spreading the fixed charges and diminishing returns from over used variable resources.

4.7.2 Benefit Cost Analysis

The benefits and costs encounter a number of measurement problems. Costs are cash outflows and benefits are cash inflows. The payback period is the cash outlays on the project to be recovered is a length of time.

$$\text{Payback period} = \frac{\text{Original cost of Investment}}{\text{Annual cash inflows}}$$

The project with the shortest payback period is selected.

1. Suitability of Method

(*i*) This method is suitable when the project has shorter gestation period and project cost is small.

(*ii*) This method is the best appraisal method when there is shortage of cash and the enterprize depends upon internal generation of cash.

(*iii*) This method is suitable when there can be rapid changes in technology.

(*iv*) This method is suitable when there is political instability.

2. Advantages of the Method

(*i*) The method is easy to operate and simple to understand.

(*ii*) The estimates are reliable and results most accurate for projects with short payback period.

(*iii*) The method is useful for concerns which are short of cash and need to cash got back in short time to invest as capital expenditure in the project.

(*iv*) The industries with high technological obsolescence donot consider cash inflows beyond payback period.

3. Disadvantages

(*i*) This method lays more emphasis on liquidity than profitability.

(*ii*) The earnings beyond payback period are not considered and projects with high rate of return after payback period may be rejected.

(*iii*) It ignores the time value of money and provides only for the recovery of the original investment.

Example 4.1. *Based on the given information, select the project most suitable based on payback period.*

Cash flows (Savings before depreciation)	*Project A (Rs.)*	*Project B (Rs.)*
Capital cost	*1,50,000*	*1,50,000*
I year	*50,000*	*40,000*
II year	*50,000*	*40,000*
III year	*50,000*	*40,000*
IV year	*20,000*	*30,000*
V year	*20,000*	*70,000*
VI year	*20,000*	*90,000*
Total cash flow	*2,40,000*	*3,10,000*
Payback period	*3 Years*	*4 Years*

Solution: The project A has payback period of 3 years and project B has 4 years. The project A is superior to project B.

4.7.3 Discounted Cash Flow Method

The investments are current capital outlays whereas the return are expected future flows. In any economy, capital has value and therefore, capital invested has time value of money. Discounted cash flow method provides a more objective basis for evaluation and selection of investment projects. This method takes care of both the magnitude and timing of expected cash flow in each period of a project life.

1. **Discounting** involves reducing the value of future return to make it directly comparable to the present value. A rupee is worth more today than the same rupee in future.
2. **Discount Rate** is the rate at which the future cash flow is reduced to the present value. The method of discounting depends upon discounting period and a suitable discounting rate.
3. **Discounting Period** is the economic life of a project and depends upon life span, technological changes, availability of raw material and market stability.

4. **Discount Rate** is the market rate of interest. Special tables, have been prepared for the present value of one rupee for different periods and discount rate.

TABLE 4.2. The Present Value of One Rupees

Year	9%	10%	11%	12%	13%	14%	15%	16%	17%	18%	19%	20%
1	0.917	0.909	0.901	0.893	0.885	0.877	0.870	0.862	0.855	0.847	0.840	0.833
2	0.842	0.826	0.812	0.797	0.783	0.769	0.756	0.743	0.731	0.718	0.706	0.694
3	0772	0.751	0.731	0.712	0.693	0.675	0.658	0.641	0.624	0.609	0.593	0.579
4	0.708	0.683	0.659	0.636	0.613	0.592	0.572	0.552	0.534	0.516	0.499	0.482
5	0.650	0.621	0.593	0.567	0.543	0.519	0.497	0.476	0.456	0.437	0.419	0.402
6	0.596	0.564	0.535	0.507	0.480	0.456	0.432	0.410	0.390	0.370	0.352	0.335
7	0.547	0.513	0.482	0.452	0.425	0.400	0.376	0.354	0.333	0.314	0.296	0.279
8	0.502	0.467	0.434	0.404	0.376	0.351	0.327	0.305	0.285	0.266	0.249	0.233
9	0.460	0.424	0.391	0.361	0.333	0.308	0.284	0.263	0.243	0.225	0.209	0.194
10	0.422	0.386	0.352	0.322	0.295	0.270	0.247	0.227	0.210	0.000	0.176	0.162
11	0.388	0.350	0.317	0.287	0.261	0.237	0.215	0.195	0.178	0.162	0.148	0.135
12	0.356	0.319	0.286	0.257	0.231	0.208	0.187	0.168	0.152	0.137	0.124	0.112
13	0.326	0.290	0.258	0.229	0.204	0.182	0.163	0.145	0.130	0.116	0.104	0.093
14	0.299	0.263	0.232	0.205	0.181	0.160	0.141	0.125	0.111	0.099	0.088	0.078
15	0.275	0.239	0.209	0.183	0.160	0.140	0.123	0.108	0.095	0.084	0.074	0.065
16	0.252	0.218	0.188	0.163	0.141	0.123	0.107	0.093	0.081	0.071	0.062	0.054
17	0.231	0.198	0.170	0.146	0.125	0.108	0.093	0080	0.069	0.060	0.052	0.045
18	0.212	0.180	0.153	0.130	0.111	0.095	0.081 .	0.069	0.059	0.051	0.044	0.038
19	0.194	0.164	0.138	0.116	0.098	0.083	0.070	0.060	0.051	0.043	0.037	0.031
20	0.178	0.149	0.124	0.104	0.087	0.073	0.061	0.051	0.043	0.037	0.031	0.026
21	0.164	0.135	0.112	0.093	0.077	0.064	0.053	0.044	0.037	0.031	0.026	0.022
22	0.150	0.123	0.101	0.083	0.068	0.056	0.046	0.038	0.032	0.026	0.022	0.018
23	0.138	0.112	0.091	0.074	0.060	0.049	0.040	0.033	0.027	0.022	0.018	0.015
24	0.126	0.102	0.082	0.066	0.053	0.043	0.035	0.028	0.023	0.019	0.015	0.013
25	0.116	0.092	0.074	0.059	0.047	0.038	0.030	0.024	0.020	0.016	0.013	0.010
30	0.075	0.057	0.044	0.033	0.026	0.020	0.015	0.012	0.009	0.007	0.005	0.004
35	0.049	0.036	0.026	0.019	0.014	0.010	0.008	0.006	0.004	0.003	0.002	0.002
40	0.032	0.022	0.015	0.011	0.008	0.005	0.004	0.003	0.002	0.001	0.001	0.001
45	0.021	0.014	0.009	0.006	0.004	0.003	0.002	0.001	0.001	0.001	0.000	0.000
50	0.013	0.009	0.005	0.003	0.002	0.001	0.001	0.001	0.000	0.000	0.000	0.000

4.7.4 Internal Rate of Return Method

Internal rate of return method is used when the cost of investment and annual cash inflows are known. This method is also called "Discounted Cash Flow Method" or "Yield Method" or "Time Adjusted Rate of Return Method".

Internal Rate of Discount equates the aggregate discounted benefits with the discounted costs.

$$\begin{bmatrix}\text{Aggregate Discounted}\\ \text{Cash Inflows}\end{bmatrix} - \begin{bmatrix}\text{Aggregate Discounted}\\ \text{Cash Outflows}\end{bmatrix} = 0$$

$$\sum_{t=0}^{t=n} \frac{B_t}{(I+1)^t} = \sum_{t=0}^{t=n} \frac{C_t}{(I+1)^t}$$

where B_t = Cash inflows during the years, 0, 1, 2, 3, ...n.

C_t = Cash outflows during the years 0, 1, 2, 3, ...n.

I = Internal rate of return

t = Time in years.

1. Method of Calculating Internal Rate of Return

(*i*) Calculate the net cash flow by subtracting cost excluding depreciation from annual sales.

(*ii*) Obtain the present value of capital cost spent in the base year.

(*iii*) Calculate the present value of net cash flow by using an appropriate rate of interest from Table 4.2.

(*iv*) Carry out the above calculations for each project and rank the preference of the project.

2. Advantages

(*i*) The method recognizes the time value of money and considers flows for the entire life of project.

(*ii*) It is expressed in percentage and has psychological appeal.

(*iii*) The discounted cost of capital is not required.

(*iv*) It fulfils the objective of maximum owner's welfare.

3. Disadvantages

(*i*) It involves complicated calculations and difficult to use in practice.

(*ii*) Decision making is very difficult as cash outflows take place at different time period.

(*iii*) It gives inconsistent results if the projects differ in their expected lives, cash flows or timing of cash flows.

4.7.5 Net Present Value Method (NPV)

This method defines the excess of present value of project cash flows over that of outflows. When all cash outflows take place in the base year.

$$\text{NPV} = \frac{B_1}{(1+r)^1} + \frac{B_2}{(1+r)^2} + \ldots\ldots + \frac{B_n}{(1+r)^n} - C_0$$

where $B_1, B_2, B_3, \ldots. B_n$ = Cash inflow in the periods 1, 2, 3,....n years

C_0 = Cash outflow in the base years, 0.

r = Cost of capital of the enterprise.

1. Decision Rules

(*i*) Accept a project which has highest Net Present Value.

(*ii*) For independent investment, accept a project if Net Present Value is more or equal to zero.

(*iii*) Reject a project with negative Net Present Value.

2. Advantages

(*i*) It indicates the value added to the total assets of the enterprize.

(*ii*) It automatically allows for the recovery of initial investment and interest for cost of investment.

(*iii*) The method recognizes the time value of money and considers all cash flows over the entire life of the project.

(*iv*) It fulfills the objective of owner's maximum welfare.

3. Disadvantages

(*i*) It is difficult to use.

(*ii*) When project involve different investments, decision becomes difficult.

(*iii*) The project with high Net Present Value may be unattractive as it may require high capital outlay.

(*iv*) It is difficult to select a proper discount rate.

(*v*) It should always be supplemented with "Benefit Cost Ratio" or "Profitability Index".

Example 4.2. *A company is considering the purchase of a new machine costing Rs. 8 lakh. The expected earnings after taxation are as follows :*

Year	*Cash flow (Rs.)*
I	*80,000*
II	*2,40,000*
III	*3,20,000*
IV	*4,80,000*
V	*3,20,000*

The company wants a return on capital of 10 percent. Find the profitability of the machine.

Solution :

Profitability Statement
(Discount factor = 10%)

Year	PV of Re 1 at 10% (Rs.)	Cash flow	PV of Cash flow
I	0.909	80,000	72,720
II	0.826	2,40,000	1,92,820
III	0.751	3,20,000	2,40,320
IV	0.683	4,80,000	3,27,840
V	0.621	3,20,000	1,78,720
	Total	14,40,000	10,37,840
	Cost	80,000	80,000
	NPV		2,37,840

The machine would be profitable since the PV of cash inflow is greater than PV of cash outflow.

Example 4.3. *An investment of Rs. 1,00,000 yields cash flows in 5 years as Rs. 25,000, Rs. 40,000, Rs. 50,000 and Rs. 30,000. The interst rate is 12%. Calculate:*

(a) Cash benefit ratio

(b) IRR

Solution :

Prepare a table for calculation of present value of cash flow. (Use Table 4.2).

Year	Cash Flow (Rs.)	Discount factor at 12%	Present value (Rs.)
0	–1,00,000	1	–1,00,000
1	25,000	0.893	22,325
2	40,000	0.797	31,880
3	50,000	0.712	35,600
4	30,000	0.636	19,080
Total	45,000		8,885

(*a*) Benefit-Cash ratio (BCR) $= \dfrac{\text{Present value of Cash inflows}}{\text{Present value of Cash outflows}}$

$= \dfrac{1{,}08{,}885}{1{,}00{,}000} = 1.08885$ **Ans.**

Net Benefit – Cahs ratio

NBCR = BCR – 1

= 1.08885 – 1

= 1.08885 **Ans.**

(*b*) The IRR is arrived at by equating the present value of investment to the present value of cash inflows. The IRR is estmated by trial and error.

(*i*) Assume, $r = 12\%$

$$\therefore \quad 1,00,000 = \frac{25,000}{(1+r)^1} + \frac{40,000}{(1+r)^2} + \frac{50,000}{(1+r)^3} + \frac{30,000}{(1+r)^4}$$

$$= 22321.4 + 31887.8 + 35589 + 19065.5$$

$$= 108865.7$$

(*ii*) Assume, $r = 13\%$

$$1,00,000 = \frac{25,000}{(1.13)^1} + \frac{40,000}{(1.13)^2} + \frac{50,000}{(1.13)^2} + \frac{30,000}{(1.13)^4}$$

$$1,00,000 = 22123.9 + 31325.87 + 346528 + 18399.56$$

$$= 106501.83$$

(*c*) Assume, $r = 15\%$

$$1,00,000 = \frac{25,000}{(1.15)^1} + \frac{40,000}{(1.15)^2} + \frac{50,000}{(1.15)^3} + \frac{30,000}{(1.15)^4}$$

$$= 21739 + 30245.75 + 32875.8 + 17152.6$$

$$= 102013$$

(*d*) Assume, $r = 16\%$

$$1,00,000 = \frac{25,000}{(1.16)^1} + \frac{40,000}{(1.16)^2} + \frac{50,000}{(1.16)^3} + \frac{30,000}{(1.16)^4}$$

$$= 21551.7 + 25000 + 32032.9 + 16568.7$$

$$= 95153.3$$

The correct value of of r must be between 15% and 16%.

$$\text{IRR}(r) = 15\% + \frac{(16-15)(102013-100000)}{(102013-95153.3)}$$

$$= 15\% \frac{2013}{6859.7}$$

$$= 15\% + 0.29345 = 15.29\%$$

Example 4.4. *For a project for which initial investment is Rs. 3,00,000 and future cahs flows are Rs. 60,000, Rs. 1,00,000, Rs. 1,20,000 and Rs. 1,50,000 for next 4 years. What is its NPV?*

Solution :

Assume a discounting factor = 15%

Year	Discounting factor at 15%	Cash flow (Rs.)	PV of Cash flow (Rs.)
0	1	–3,00,000	–3,00,000
1	0.870	60,000	52,200
2	0.756	1,00,000	75,600
3	0.658	1,20,000	78,960
4	0.572	1,50,000	85800
	Total PV		292560
	NPV		–7440

The project is not viable since pV of cash inflows (Rs. 2,92,560) is less than pV of outflows. (Rs. 3,00,000).

Question Bank

1. Explain the importance of a business project report. What precaution should be observed while preparing a report?
2. What are the main contents of a project report? Explain.
3. What do you mean by a "Feasibility Report" and what are the main contents?
4. Explain the difference between "Feasibility Report" and "Project Report."
5. What do you understand by project appraisal? Discuss project appraisal methods.
6. What is a "Preliminary Project Report (PPR)? What are its contents and uses. Design a proforma of a Preliminary Project Report?
7. What is Techno-economic feasibility report? What aspects are analysed?
8. Design the proforma of a detailed project report.
9. How is project viability established.
10. Write a brief note on demand analysis.
11. Explain the following :
    (*a*) Material Balance and Output Method.
    (*b*) Benefit Cost Analysis
12. Compare the advantages and disadvantages of Internal Rate of Return Method and Net Present Value Method.
13. What is the significance of Discounted Cash Flow Method?

PART – III

Accountancy and Management of Small Business

Chapter

5

Accountancy

5.1 INTRODUCTION

An entrepreneur who is setting up an industry or business should form a system of maintaining books of accounts from the very beginning. This system is called accountancy. Inadequate system of maintaining account leads to failure of business. Accounts show the economic condition of any business or industry. Accounts can be defined as a 'mirror' of the trade or industry.

5.2 PURPOSE OF ACCOUNTANCY

The accountancy can be effectively used for the following purposes :

1. To ascertain whether the business is making profit or loss.
2. To get information about accounts payable to others and amounts receivable from others.
3. To control the expenditure by companying income and expenditure with those of previous year.
4. To ascertain the financial position of business.
5. To plan the business accurately.
6. To meet legal requirements of income tax, sales tax, excise duty and labour legislation.
7. To estimate cost of the products.
8. To calculate the rate of return from business.
9. To obtain readily the amount of investment and type of capital investment in business.
10. To get comparative picture of the business with respect to other similar business.
11. To avoid fund shortfalls.
12. To develop qualities of accuracy and methodology.

5.3 REQUIREMENTS

The following steps must be taken to take full advantage from accounts.

1. A good accounting system should be designed with the help of an expert.
2. A trained accountant should be employed.
3. The accounts must be written regularly.
4. The accounts must be supervised periodically and frequently.
5. Regular and accurate information must be given.

There are two methods of accountancy used in India.

1. Traditional system (Deshi Nama System) where accounts are written in regional language.
2. Double Entry Book Keeping.

5.4 PROCEDURE

The following should be followed carefully.

1. Maintain account books, statements and other required stationery.
2. Record all entries on daily basis.
3. Accountant must submit the collection list everyday and before leaving the office.
4. List of creditors should be submitted regularly at least one week in advance.
5. List of major payments should be submitted at least 15 days in advance.
6. Trial Balance, Trading Account and Profit and Loss Account should be prepared by the end of every month.
7. Cash flow statement for the next six months should always be kept ready in advance.
8. Bank Balances should be matched periodically with the books.
9. Tax consultant should be contacted frequency and the latest information of the business should be given to him.
10. Regularly meet the head of financial institution and lending bank and keep them informed about the progress and future plans.
11. Daily receive the figures of income, expenditure, purchases and sales.
12. Compare the actual expenditure with the estimated amount.
13. All bills, receipts, vouchers should be prepared with full details.
14. For purchase of fixed assets, check the availability of surplus funds or necessity for raising loans.
15. The fixed assets and stock of goods should be recorded in the stock register.
16. Required return form should be despatched to concerned banks, government and other statutory authorities in time.
17. The payable amounts should always be paid before due dates. Creditors should be informed if timely payments are not possible.

18. "Post-dated" cheques should never be used and no check should be dishonoured.
19. Accounts for every praise must be tallied.
20. Receipts must be prepared immediately on receiving payments.
21. Payments must be sanctioned by authorized person, proper vouchers must be prepared for every payment.

A systematic and regular accounts keeping leads to the following advantages.

1. Shortfalls of funds can be prevented.
2. Undue tension can be avoided.
3. Reputation in society can be enhanced.
4. Success can become a reality.

5.5 OFFICE OF ACCOUNTANT

The accountant is the person who maintains accounts. He has to maintain certain books.

5.5.1 Qualities of an Accountant

1. Knowledge of accounts
2. Honesty
3. Good handwriting
4. Ability to work for long hours
5. Accuracy
6. Regularity

5.5.2 Books to be Maintained

1. Cash Book with Bank Column
2. Ledger
3. Purchase Register
4. Sales Register
5. Journal Proper
6. Salary Register
7. Stock Register
8. Dead Stock Register

5.5.3 Stationery for Accounts

1. Receipt Book
2. Cash Memo Book
3. Bill Book
4. Voucher Book
5. Delivery Challan Book.

5.6 CLASSIFICATION OF ACCOUNTS

Financial accounting is systematic recording of monetary impact of transaction. There are two types of income and two types of expenditure.

1. Capital Income is related to disposal of fixed assets like land, buildings, machinery, etc.
2. Revenue income is related to sales of current assets during current year.
3. Capital Expenditure is related to acquisition of fixed assets.
4. Revenue Expenditure is related to purchase of current assets and factory purchases.

5.6.1 Double Entry Book Keeping

Each transaction is referred as debit and credit and noted twice once as debit and second time as credit. This system of recording is known as Double Entry Book Keeping. The debit and credit are decided as per following rules.

Transaction	Debit	Credit
Purchase	Purchase Account (Material comes in)	Suppliers Account (supplier gives)
Sales	Customer Account (Customer receives)	Sales Account (Material goes out)
Payment by cash/cheque	Receiver (Supplier) Expenditure	Cash (Cash transaction) Bank (Cheque transaction)
Receipt by cash/cheque	Cash (Cash transaction) Bank (Cheque transaction)	Debtor's Account (Givers Account)

5.6.2 Groups of Accounts

There are two major groups of accounts

1. Asset and Liability Account called Balance Sheet.
2. Income and Expenditure Account called Profit and Loss Account.

(1) Balance Sheet as on .... (Date)

Liabilities		Assets	
Code Group	Account Group	Code Group	Account Group
10	Creditors (Goods received but payments not made)	20	Debtors (Goods sold but payments not received)
11	Own funds	21	Fixed assets (land, building machinery etc.)
12	Borrowed funds	22	Investments (in shares of other companies)
13	Sundry liabilities (for expenses, etc.)	23	Closing stocks
		24	Loans, Advances and Deposits
		25	Cash and Bank Balances.

(2) Profit and Loss Account for the.... (Year)

Expenditure		Income	
Code Group	**Account Group**	**Code Group**	**Account Group**
30	Opening stocks	41	Sales
31	Purchases		
32	Salaries, wages and fringe benefits	42	Other Income
33	Production expenses	22	Closing stocks
34	Administration expenses	11	Losses
35	Selling expenses		
36	Distribution expenses		
37	Interest		
38	Depreciation		
39	Taxes		
40	Profits		

Note : The code number is for convenience of reference.

5.7. ASSESSMENT OF ECONOMIC VIABILITY

The project viability as given below should be part of the detailed project report.

1. **Profitability Calculations**

Estimated profitability and financial ratios are given for three years.

S. No.	Description	1st Year	2nd Year	3rd Year
1.	Sales Revenue			
2.	Estimated Production Expenses			
3.	Production Profitability (1 – 2)			
4.	Other income			
5.	Profit before Tax (3 + 4)			
6.	Expected income tax to be paid			
7.	Net profit (5 – 6)			
8.	Net cash returns (7 + depreciations + investment allowance)			
9.	Percentage of net profit (*i*) On sales $\left|\dfrac{(7)}{(1)\times 100}\right|$ (*ii*) On investment			

2. **Break-even Point (Calculation and Explanation)**

$$\% \text{ of Breakeven} = \frac{\text{Fixed Expenses}}{\text{Expected Contribution}} \times 100$$

where Contribution = Income from sales – Variable Expenses.

3. **Cash Flow**

S. No.	Description	1st Year	2nd Year	3rd Year
(*a*)	**Financial Inflow**			
1.	Net profit (income tax and interest added but depreciation substrated)			
2.	Own investment			
3.	Depreciation			
4.	Investment allowance			
5.	Long-term loan			
6.	Short-term loan			
7.	Other income			
(*b*)	**Financial Outgo**			
1.	Preliminary and pre-operative expenses			
2.	Fixed investment			
3.	Increase in current assets			
4.	Reduction in long-term loan			
5.	Reduction in short-term loan			
6.	Interest			
7.	Income tax			
8.	Total			
(*c*)	**Repayment of Loans**			
1.	Net profit (income tax debited)			
2.	Entrepreneurs drawings			
3.	Available profit (1 – 2)			
4.	Amount to be returned to financial institutions			
5.	Interest on funds availed from state corporation and banks			

4. **Project Implementation Planning (Time in Weeks)**

(*i*) Preliminary survey of product.
(*ii*) Market research and preparation of project report.
(*iii*) Loan application, preparation and sanction.
(*iv*) Selection of site and other necessary formalities.

(*v*) Order for plant and machinery.
(*vi*) Construction of factory building (rented building).
(*vii*) Installation of plant and machinery.
(*viii*) Selection of personnel.
(*ix*) Application and formalities for electricity, fuel.
(*x*) Other necessary licenses and sanction letters necessary for project implementation.
(*xi*) Electrification in factory.
(*xii*) Trial production.
(*xiii*) Commercial production.
(*xiv*) Training of personnel.
(*xv*) Inauguration of factory.

Prepare bar chart, procedence diagram or PERT–CPM charts.

5. **Enclosures**
   1. Education certificates
   2. Experience certificates
   3. Testimonials for licence and permission letters.
   4. Drawings of factory building and expected cost of building or rent receipt.
   5. Quotation of machinery and equipment (in triplicate)
   6. Letters from raw materials suppliers (if possible)
   7. Partnership deed or any other documents.

6. **Financial Features at a Glance**

S. No.	Description	Own Investment	Long-term loan	Short-term Loan from Bank	Total
1.	Land and Building				
2.	Plant and Machinery				
3.	Working Capital				
4.	Preliminary and pre-operative expenses				
5.	Total				
6.	Expected total production expenses				
7.	Profitability				
8.	Expected net profit (less income tax)				
9.	Break even point (%age)				

7. **Precautions**

(*i*) Always estimate cost on high side and income on lower side.
(*ii*) Use only personally verified costs and prices.

(*iii*) Data taken from other project report must be updated.
(*iv*) Make escalation provisions due to planning period and implementation period.
(*v*) Take 50 to 60% capacity utilization in 1st year, 80% in 2nd year.
(*vi*) Project report should be prepared personally.
(*vii*) Counter check all calculation.
(*viii*) Take advice and opinion of expects.

8. **Reasons of Failure of Project Report**

(*i*) Over estimation of capacity utilization.
(*ii*) Under-estimation of provisions for contingencies.
(*iii*) Under-estimation of cushion for fixtures and electrification
(*iv*) Under-estimation of cushion for handling and time loss
(*v*) Unrealistic report to present on rosy picture.
(*vi*) Without updated and relevant information and documents.
(*vii*) Overestimation of income and under-estimation of expenses.

5.8 DECISION MAKING

Decision-making is a process of selection of a course of action from a number of possible alternatives. The selected alternative should more satisfy the objectives of decision problem than others in a given situation.

5.8.1 Characteristics of Decision-Making

Following are the basic characteristics of decision-making.

1. It is a process of selection of a course of action from among alternative courses of action.
2. It is process of application of human intellectual abilities.
3. It is an end process achieved by deliberation and reasoning.
4. It is situation specific controlled by environment.
5. It involves a time dimension and time lag.
6. It is a function of purpose.
7. It involves defining a problem and analysing various alternatives and selecting most satisfying alternative.

5.8.2 Types of Decision

Decision may be classified as follows :

1. Tactical and Routine Decisions

Such decisions are made repeatedly by managers following some established rules, procedures and policies. Such decisions are related to policy matters and require through fact finding and analysis of possible alternatives.

2. Policy and Operating Decisions

Policy decisions are taken by top management and operating decisions are taken by lower management to put policy decisions into action.

3. Organizational and Personal Decisions

A manager takes organizational decision in his official capacity and delegates the same. A manager takes personal decision in his individual capacity not related to organization.

4. Progammed and Non-programmed Decisions

Programmed decisions are routine decisions following a standard procedure and may concern one employee. Non-programmed decision may concern a large number of employees and require thorough analysis of situational factors.

5. Individual and Group Decisions

Individual decision is taken by one person in a small organization. Group or collective decisions are taken in large organization by a group of members.

5.8.3 Procedure of Decision-Making

The following stages are involved in scientific decision-making.

1. Definition of the problem
2. Analysis of the problem
3. Data collection
4. Development of alternatives
5. Review of key factors
6. Selection of best alternative
7. Implementation of decision
8. Follow-up action

1. Definition of the Problem

Clear definition of the problem is very important before finding right solution. Sufficient time should be spent to define the problem causing the trouble and requiring a correction.

2. Analysis of the Problem

The analysis of the problem needs classification of the problem so that the level of management to take decision and the type of consultation required can be specified.

3. Data Collection

Collection of proper data and information is needed for taking correct decision.

4. Development of Alternatives

The principle of limiting factor which is obstructing the desired goal should be used in the development of alternatives.

5. Review of Key Factors

The key factors may be inadequate funds, shortage of human resources, old machine or lack of marketing skills.

6. Selection of Best Alternative

Final selection of alternative requires evaluation of all alternatives against the following consequences.

(*i*) Degree of risk involved

(*ii*) Economy of effort

(*iii*) Timing of decision

(*iv*) Limitation of resources.

7. Implementation of Decisions

Systematic steps should be taken to implement the decision taken by educating and informing the persons who may be affected by the action. The principle of slow and steady progress may be followed to bring about the changes in the behaviour of the affected employees.

8. Follow-up Action

There should be an efficient system of feedback information. A good decision can be a model for similar future problems. A bad decision should not be repeated in future.

5.9 EXPECTED COSTS

Costing is a process of classifying, recovering and appropriately allocating expenditure for determination of cost of the product or services. Costing is the process of determination of cost of production and selling a product or service. Proper record and system of costing has to be adopted for controlling the cost.

5.9.1. Functions of Cost Estimation

After the capital is invested, procedures are established to find the financial efficiency of the project. Cost accounting uses the accounting and costing principles, methods and techniques in the ascertainment of costs and the analysis of savings.

Objectives of cost accounting :

1. Determination of actual cost of the product.
2. Provide basis for pricing the product.
3. Control of cost.
4. Guide for making operating (business) decisions.

The functions of cost estimation are :

1. Project Costing and Feasibility Analysis

It will give the finance requirement and economic soundness of the project.

2. Cost of Product or Service

The final cost of a product or service includes :

(*i*) Cost of production.

(*ii*) Cost of selling and distribution.

(*iii*) Cost of after-sales service.

3. Product Design

The estimation of cost of material gives guidance for product design improvement to make it more economical.

4. Production Cycle

The estimation of machine hours, labour hours, etc. help to estimate the production time and time of delivery.

5. Overhead Expenses

The estimation of overhead expenses help to classify various expenses like packing, transportation, rent, watch and ward and other administrative expenses and guides to reduce and control the expenditure under different heads.

6. Make a Busy Decision

The product cost estimation also helps to take decision whether to manufacture the component in the plant or buy from the market.

7. Standard for Comparison

Cost estimation gives a basic standard against which actual expenditure can be calculated.

5.9.2. Cost Estimation Procedure

Cost estimation of a project is carried out on the basis of market survey and research where past record experience of other manufactures, distributors, customers and consultants are used. The data is collected which is subject to mathematical analysis. However, when a new product is launched, cost estimation is based on calculations of direct and indirect material cost, labour cost, overheads, distribution cost, etc.

The procedure of product cost estimation is discussed below :

1. Bill of Materials

Prepare bill of materials including rolled material, castings and forgings and other boughts out items.

2. Make or Buy Decision

Separate the parts to be made in the plant or to be bought from the market on the basis of economic feasibility report.

3. List of Tools, Equipment, Jigs and Fixtures

Prepare a list of machines, jigs, fixtures, tools needed for the production.

4. Process Sheet

Prepare the process sheets giving the sequence of various operations.

5. Processing Data

Determine the machining data on cutting speeds, depth of cuts and feeds for all operations.

6. Machining Time

Evaluate the machining time for all parts.

7. Machining Allowance

Add tool handling time, tool setup time, machine adjustment time, down time allowances to the machining time to work out total time.

8. Material Cost

Estimate material cost from bill of materials.

9. Labour Cost

Estimate labour cost from the total machining time.

10. Prime cost

Add labour cost to the material cost to get prime cost.

11. Overhead Charges

Estimate overhead charges based on factory norms.

12. Total Cost

Add up prime cost and overheads to get the total cost.

13. Sale Price

Add distribution cost and profit to fix the sale price.

14. Market Price

Add discount to be given to the distributors to calculate the market price.

15. Management Approval

Get the final approval of the management on the cost structure.

5.9.3. Elements of Cost and Ladder of Cost

Elements of the cost of an item produced in a factory are shown in the ladder of cost illustrated in Fig. 5.1.

1. Material Cost

It consists of cost of materials directly used in the product. The direct material cost includes raw material cost and cost of bought-out items. The materials indirectly used like electrodes, lubricating oil, tools, etc. are indirect materials and are counted as factory overheads.

2. Labour Cost

It consists of direct labour who actually process the product. Wages and salaries of those who are not directly concerned with the fabrication process like helpers, supervisors, etc., are counted as factory overheads.

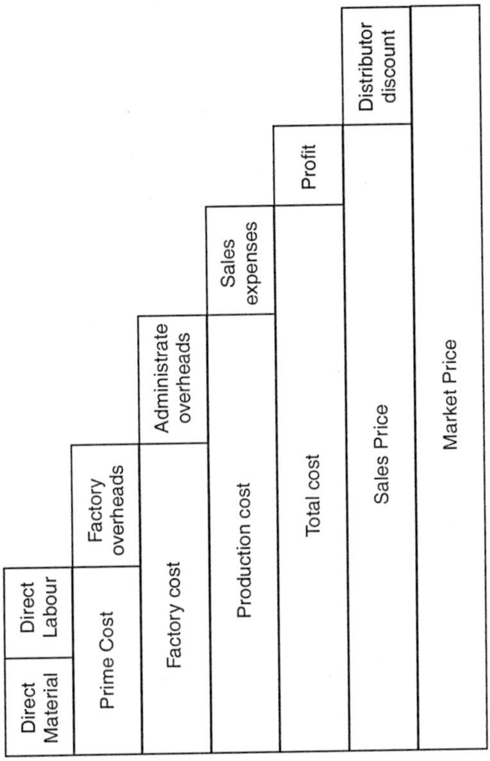

Fig. 5.1. Ladder of cost.

3. Prime Cost

Prime cost is the sum of cost of direct materials and direct labour.

4. Overheads (Indirect Costs)

Overheads are indirect expenses and are classified as :

(*i*) Factory overheads.

(*ii*) Administrative overheads.

(*iii*) Sales overheads.

5. Total Cost

The total cost of a product consists of the sum of prime cost and factory overheads.

Cost of social security, pensions and health and accident insurance which are approximately proportional to direct labour costs are included in the direct labour cost of the product. Earlier, these were taken as indirect cost and included in overheads.

6. Market Price

The distributor's discount and profit are added to total cost to work out the market price of the product.

5.9.4. Depreciation

5.9.4.1. Concept

Depreciation is a gradual shrinkage in the value of a fixed asset. It is also the process of allocating the cost of an asset (less salvage value) in a logical manner over the estimated life of the asset.

The causes of depreciation are :

1. Depreciation Due to Physical Condition

(*a*) Wear and tear during use.

(*b*) Decay due to rust, heat, etc.

(*c*) Poor maintenance.

(*d*) Accident.

2. Depreciation Due to Functional Features

(*a*) Inadequacy

(*b*) Obsolescence

Depreciation is the result of the use of assets, passage of time and obsolescence.

3. Obsolescence

Certain assets become obsolete due to the following:

(*a*) Customer taste/fashions/demand suddenly falls

(*b*) New equipment in the market capable of producing more products of good quality with less labour and processing time, makes old equipment obsolete due to tough competition in the market.

4. Service Life of Assets

Service life of an asset is the period during with the asset gives optimum performance without any trouble.

5. Factors Affecting Depreciation

The depreciating period depends upon:

1. Acquisition cost
2. Estimated life.
3. Estimated salvage value.
4. Selection of depreciation method.

5.9.4.2. Methods of Charging Depreciation

The following methods are commonly used for charging depreciation :

1. Straight Line Method.
2. Diminishing Balance Method.
3. Sum-of-years Digit Method.
4. Sinking Fund Method.
5. Annuity Charging Method.
6. Machine Hour Basis of Charging Depreciation.

1. Straight Line Method

This method provides equal periodic amount to be charged yearly over the useful life of the asset as depreciation. It assumes that the asset depreciates at a constant rate on its economic life. The annual depreciation amount is :

$$D = \frac{C-S}{N} \quad \text{[Rs]}$$

where C = Initial cost of machine + Installation + Repair [Rs.]

S = Scrap value at the end of its useful life [Rs.]

N = Number of years of useful life.

Advantages :

1. The method is simple and easy.
2. Calculations are easy.
3. It recognizes that the decline in value of asset is directly proportional to its age.

Limitations :

1. Method is unrealistic since assets do not depreciate uniformly.
2. Repair and maintenance increase with passage of time.

Example 5.1. *Calculate the yearly depreciation for the lathe using straight-line method of depreciation. The details are given below :*

Initial cost of lathe = Rs. 60,000

Installation charges = Rs. 2,000

Repair and maintenance charges = Rs. 3,000

Scrap value = Rs. 6,000

Useful life of lathe = 10 years.

Solution :

$$C = 60{,}000 + 2{,}000 + 3{,}000 = \text{Rs. } 65{,}000$$

$$S = \text{Rs. } 6{,}000$$

$$N = 10 \text{ years}$$

$$D = \frac{C-S}{N} = \frac{65{,}000 - 6{,}000}{10} = \text{Rs. } 5{,}900$$

The depreciation will be charged for 10 years at a uniform rate of Rs. 5,900 every year.

2. Diminishing Balance Return

It is assumed that the machines and equipments depreciation at a faster rate in the beginning of their life span. The depreciation fund should also be more during the early years when repairs and renewals are not costly. Equipment book value goes on decreasing with time. Certain percentage of the current book value is charged as depreciation.

Depreciation factor,

$$X = 1 - \left(\frac{S}{C}\right)^{\frac{1}{N}}$$

where X = fixed percentage for calculating annual depreciation.
C = Initial cost of installed equipment.
S = Salvage value.
N = Estimated equipment life.

The depreciation amount at the end of first year

$$= XC$$

The book value at the end of first year

$$= C - XC$$
$$= C(1 - X)$$

The depreciation amount at the end of second year

$$= XC(1 - X)$$

The book value at the end of second year

$$= C(1 - X) - XC\ (1 - X)$$
$$= C(1 - X)^2$$

Similarly book value at the end of N years

$$= C(1 - X)^N$$

or,
$$S = C(1 - X)^N$$

Solving the equation,

$$X = 1 - \left(\frac{S}{C}\right)^{\frac{1}{N}}$$

The depreciation amount at the end of N years.

$$= CX(1 - X)^{N-1}$$

Example 5.2. *A machine is purchased for Rs. 75,000 with an estimated age of 10 years and scrap value of Rs. 5,000. What will be the depreciation for the 6th year and its book value at the end of 6th year? Use declining balance method of depreciation.*

Solution :

$$C = \text{Rs. } 75{,}000$$
$$S = \text{Rs. } 5{,}000$$
$$N = 10 \text{ years}$$

Depreciation at the end of the 6th year

$$D = CX(1 - X)^{6-1}$$
$$= 75{,}000\ X(1 - X)^5$$

$$X = 1 - \left(\frac{S}{C}\right)^{\frac{1}{N}} = 1 - \left(\frac{5000}{75000}\right)^{\frac{1}{10}}$$
$$= 0.2372$$

$$\therefore \quad D = 75{,}000 \times 0.2372\ (1 - 0.2372)^{1/5}$$
$$= \text{Rs. } 4{,}593$$

$$\text{Book value} = C(1 - X)^6$$
$$= 75{,}000\ (1 - 0.2372)^6$$
$$= \text{Rs. } 14{,}770$$

3. Sinking Fund Method

Sinking fund is established to acculate fund for replacing the equipment at a proper time when it becomes uneconomical to continue using it. The fund equals the loss is value of the asset taking into account an interest i on the fund so collected. An identical rate of depreciation is charged every year throughout the life of the asset.

Annual rate of depreciation,

$$d = \frac{i(C-S)}{(1+i)^N - 1}$$
$$= (C - S)\left(\frac{A}{F}, i, N\right)$$

Depreciation upto age n years,

$$D_n = (C - S)\frac{\left(\frac{A}{F}, i, N\right)}{\left(\frac{A}{F}, i, N\right)}$$

Example 5.3. *An industrial plant with initial value of Rs. 30,000 and a salvage value of Rs. 3,000 at the end of 10 years is sold for Rs. 20,000 after one year. What is the profit or loss if sinking fund method at 8% interest rate compounded annually is used.*

Solution : Rate of depreciation,

$$d = \frac{i(C-S)}{(1+i)^N - 1}$$
$$= \frac{0.08(30{,}000 - 3{,}000)}{(1.08)^{10} - 1}$$
$$= 1864.$$

Depreciation fund at the end of one year

$$D_1 = 1 \times (1864)$$
$$= \text{Rs. } 1864$$

Book value at the end of one year

$$= 30{,}000 - 1{,}864$$
$$= \text{Rs. } 28{,}136$$

Plant sold for Rs. 20,000

$$\text{Profit/loss} = \text{Book value} - \text{Selling price}$$
$$= 28{,}136 - 20{,}000$$
$$= \text{Rs. } 8{,}136\ (\text{loss})$$

5.9.5. Overhead Expenses

Indirect expenses are called overhead expenses. The main groups of overheads are :

1. Factory Overheads

Repair and maintenance of plant and machinery, building, insurance of factory godowns, telephone bills, rent, power bills, interest on loan.

2. Administration Overheads

Salaries of administrative staff, stationery and printing, office expenses, legal expenses.

3. Selling and Distribution Overheads

Salaries to sales manager and staff, travelling expenses, advertising expenses, showroom expenses, discounts to customers, entertainment expenses etc.

5.9.5.1. Classification of Overhead Expenses

1. Fixed Overheads

The expenses which do not change with volume of production.

(*i*) Building rent.

(*ii*) Depreciation of fixed assets.

(*iii*) Insurance charges of machines and buildings.

(*iv*) Salary of managers and office staff.

(*v*) Interest.

2. Variable Overheads

The expenses which change with volume of production.

(*i*) Cost of power.

(*ii*) Repair and maintenance expenses.

(*iii*) Cost of consumable material.

5.9.5.2. Distribution of Overhead Expenses

The methods of distribution of overhead expenses are :

1. Percentage of other costs.
2. Ratio of man hour or machine hour.
3. Ratio of output.

1. Percentage of Other Costs

(*i*) *Percentage of Prime Cost*

$$\text{Overhead rate (\%)} = \frac{\text{Total overhead expenses}}{\text{Total prime cost}} \times 100$$

(*ii*) *Percentage of Direct Labour*

$$\text{Rate of overheads (\%)} = \frac{\text{Total overhead expenses}}{\text{Total direct labour cost}} \times 100$$

(*iii*) *Percentage of Direct Material*

$$\text{Rate of overheads (\%)} = \frac{\text{Total overhead expenses}}{\text{Total direct material cost}} \times 100$$

2. Ratio of Man/Machine hour

(*i*) $$\text{Man-hour rate} = \frac{\text{Total overhead expenses}}{\text{Total production man hour}}$$

(*ii*) $$\text{Machine-hour rate} = \frac{\text{Total overhead expenses}}{\text{Total machine hours}}$$

3. Unit Rate

$$\text{Unit Rate} = \frac{\text{Total overhead expenses}}{\text{Total output}}$$

Example 5.4. *Factory overheads for an organization for the year 2005-06 were Rs. 8,000 and total direct wages paid to workers during the year were Rs. 32,000. Find the %age on cost on direct labout cost.*

Solution : $$\text{Percentage oncost} = \frac{\text{Total overhead expenses}}{\text{Total direct labour cost}} \times 100$$

$$= \frac{8,000}{32,000} \times 100$$

$$= 25\%$$

Example 5.5. *A manufacturing concern produces 5,000 machines per year. The overhead expenses for the year were Rs. 8,00,000. Calculate the unit rate of overhead.*

Solution : $$\text{Unit rate} = \frac{\text{Total overhead expenses}}{\text{Quantity of production}}$$

$$= \frac{8,00,000}{5,000}$$

$$= \text{Rs. 160 per machine}$$

5.9.5.3. Cost Control

Cost control is an important function of management and is the continuing process of measuring actual results in relation to the results which were planned. The aims of cost control are :

1. Costs should not exceed certain limit.
2. Measures should be adopted to reduce the cost.

The cost control procedure has four steps.

1. **Establishment of Standards :** This is done by budgeting and standard costing.
2. **Measurement of Performance :** By cost accounting.
3. **Comparison :** By cost analysis or financial ratio analysis.

4. **Remedial Action :** By implementation of cost control and cost reduction.

5.9.6. Control of Operating Costs

In order to earn good profit, each and every cost should be kept under control. The elements of cost are :

(*i*) Control on prime cost.

(*ii*) Control on overheads.

(*iii*) Control on indirect materials and tools.

1. Control on Prime Cost

The prime cost consists of

(*i*) Direct material cost

(*ii*) Direct labour cost.

A. Controlling Material Cost

It is necessary to pay maximum attention to controlling material cost which may be 20% to 70% of product cost.

(1) ***Control Over Abnormal losses***

There are losses due to careless handling, theft, fire, and damage, etc.

(2) ***Material Handling Charges***

The materials must be stored, transported and handled in a planned way to minimize handling charges.

(3) ***Control Over Inventory***

Modern techniques must be adopted to control inventory.

(4) ***Control Over Indirect Materials***

Norms of consumption should be prepared for consumption of fuels and lubricants.

(5) ***Reclamation of Loss from Defective products***

Defective products should be dismantled.

(*i*) Serviceable parts should be reused.

(*ii*) Defective parts should be rectified and used.

(*iii*) Completely spoiled parts should be sent to foundry or sold as scrap.

6. ***Control Over Consumption in the Workshop***

This can be done by comparison with standards set.

B. Control of Labour Cost

1. Comparison of performance with standard time set by work study.

2. Reduction of time by standardization and simplification of process.
3. Reduce idle labour time by effective maintenance and prompt repairs.
4. Simplification of product design by value engineering.
5. Reduction of manufacturing time by proper use of tools, jigs and fixtures.
6. Adopt network techniques for monitoring.

2. Control on Overheads

It is more effective to reduce overheads rather than prime cost.

1. Total overhead charges of different departments should be estimated by scientific method.
2. Control indirect labour force.
3. Avoid extra time.

A. Control on Indirect Materials and Tools

The cost should be kept under control by comparison with standard costs worked out.

Methods of Controlling Overheads Costs

(1) *Budgets*

(*i*) Prepare budgets for overhead expenses for each department.
(*ii*) Compare actual expenses with budgets.
(*iii*) Analyze differences.
(*iv*) Take remedial action.

(2) *Cost Consciousness*

Education and training of people about cost consciousness, avoiding wastages of resources, efforts, etc. can be very effective.

(3) *Value Engineering*

Cost reduction through changes in product design and modification in use of material used is an effective total of value engineering.

Techniques of Cost Control

1. Standard costing.
2. Inventory control.
3. Quality control.
4. Production control.
5. Value analysis.
6. Work study.
7. Budget and budgeting control.

5.9.7. Control of Capital Cost

The capital cost is invested in the plant, equipment, land and building, etc., which cannot be disposed off without breaking the business. The fixed capital is required for :

1. Land and land development.
2. Main building and auxiliary buildings.
3. Plant and machinery for production.
4. Foundations and installations.
5. Services : Power supply, water supply, steam and air supply, drainage system, pollution control system.
6. Tools, jigs and fixtures.
7. Office equipment.
8. Furniture.
9. Patents, trade marks and pre-operative expenses.

The cost of plant and machinery will include :

1. Ex-works cost.
2. Transportation charges.
3. Handling charges : Loading and unloading.
4. Insurance during transportation.
5. Taxes : Excise duty, sales tax, octroi.
6. Foundation.
7. Erection and commissioning charges.
8. Cost of spares for two years operation.

20% of ex-works cost may be added for rough estimation.

The control of capital cost required for purchase of fixed assets can be through planning and scheduling. The cost can be controlled by timing of placement of orders and controlling the delivery period through expediting. The networks or bar charts or precedence network diagrams can be used to control the above factors.

5.10. PREPARATION OF FINANCIAL REPORTS

Financial reports are prepared periodically. These reports are used for the following purposes.

1. To appreciate the progress of the enterprize for a given period.
2. To plan and control the business operation in future.
3. To apprize the creditors share holders and employees of the enterprize with the results achieved during a given period.
4. The main financial reports are :
   (i) Trading Account
   (ii) Profit and Loss Account
   (iii) Balance Sheet
   (iv) Miscellaneous

5.10.1 Trading Account

The main features of this report are as follows :

1. It shows the results of buying and selling of goods during a given period.
2. It is prepared by matching the selling prices of the goods and services rendered with the cost of goods and services.
3. It is prepared to find out the gross profit and loss of the enterprize during the accounting period.
4. The operating stock, net purchases and expenses directly connected with the purchases are shown as debit.
5. The net sales and closing stocks are shown as credit.
6. The closing entries are passed to transfer balance of all the items due to approve therein.
7. Gross profit or gross loss shown in the Trading Account is transferred to Profit and Loss Account.

5.10.2 Profit and Loss Account

The main features of this report are as follows :

1. It highlights the net profit or net loss incured by the enterprize for a given period of business operation.
2. This report is needed to know the net results of business operation for a given period.
3. The report is analysed to control the expenses of business operation of the enterprize.
4. The losses and revenue expenses are shown as debit.
5. The gains and gross profit are shown as credit.
6. Closing entries are passed to transfer balance of accounts of items of expenses and gain.
7. The net profit or loss shown by Profit and Loss Account is transferred to the Capital Account.

5.10.3 Balance Sheet

The main features of balance sheet are as follows :

1. It is a statement of liabilities and assets of a business enterprize and shows the financial position on a given date.
2. The information given in a balance sheet is true and valid only on a given date usually 31st March of a year.
3. The balance sheet is prepared to show the financial position of a business enterprize in terms of values of assets and liabilities.
4. Balance sheet is not an account but a statement and form part of final accounts.

5. The total values of assets and liabilities are always equal, and poryrays the Accountancy Equations.
6. Assets and liabilities are marshalled in order of liquidity or order of performance.

5.11 STUDY OF ACCOUNTS AND STORES

The accounts of an enterprize are studied by auditing. The stores are studied with the help of stock register.

5.11.1 Audit of Accounts

It covers the verification of accounts and whether accountancy procedure has been strictly followed. It is compulsory and essential to find out the accuracy and authoritatively of the records maintained by the organization. It helps the management to reduce and minimize the expenses and losses.

1. Scope of Audit

It is compulsory is India as per Indian Companies Act, 1956. The accounts must be maintained by the enterprize for every year though audit may not be compulsory. Cent per cent checking of all cost transactions and accounts may not be possible and auditor may do selective checking of accounts.

2. Objectives of Audit

The objectives of audit of account are as follows :

(*i*) Audit helps to ensure proper maintenance of accounting records by the enterprize.

(*ii*) A true and fair account is ensured.

(*iii*) It helps to make a comparison of actual cost with the standard cost, find the variance and make suggestions for improvement.

(*iv*) Audit helps the management to take proper decision of the pricing of goods and services.

(*v*) Audit helps the management to reduce cost and expenses by increasing operating efficiency, productivity and reducing wastage.

(*vi*) It helps the management to ensure optimum utilization of resources.

(*vii*) It helps the management to take appropriate decision based on the data of accounts.

(*viii*) Audit is useful to customers and society as it checks the cost to lower the prices.

(*ix*) Audit helps the true and fair valuation of inventories in the Profit and Loss Account and Balance Sheet.

(*x*) It helps and provides accurate and reliable facts and data to the management.

5.11.2 Stock-taking of Stores

Study of stores is the physical verification of stock of materials in the stores. The main features of study of stores are as follows :

(*i*) The stock taking should be carried out on day-to-day basis.
(*ii*) It helps a better control over the stock of the materials.
(*iii*) The physical verification of stock of materials should be carried out by a person not connected with the store.
(*iv*) The physical balance of material should match with the balance shown in bin card as well as store ledger.
(*v*) The store keeper must give the explanation for any discrepancy. Adjustments must be made in the bin card and stores ledger accordingly.
(*vi*) The entries in the bin card and store ledger must be compared with GRN and MIR.
(*vii*) The stock taking must be carried out atleast once a year.
(*viii*) For smooth management of stores, the storekeeper himself should check the stores once in a period of three to four months.

Example 5.6. *Prepare a Balance Sheet and Profit and Loss account for a compaxy from the data given:*

Cost of goods sold	*Rs. 1,20,000.00*
Sales	*Rs. 2,00,000.00*
Operating Expenses	*Rs. 75,000.00*
Total Assets	*Rs. 1,60,000.00*
Total Liabilities	*Rs. 70,000.00*
Net Owner's Equity	*Rs. 90,000.00*

Solution :

Profit and Loss Account

Description	(Rupees)
1. Income	
Sales	2,00,000
2. Expenditure	
Cost of goods sold	1,20,000
Operating expenses	75,000
3. Gross Profit	5,000
4. Balance carried forward	5,000

Balance Sheet

Liabilites	Rs.	Assets	Rs.
1. Total Liabilities	70,000	Total assets	1,60,000
2. Net Owner's Equity	90,000		
	1,60,000		1,60,000

Question Bank

1. What is accountancy? What is the purpose and requirements of accountancy in a small business?
2. What are the important books maintained by an accountant? What procedure has to be followed in accountancy?
3. What are different types of accounts? Explain how a balance sheet is prepared.
4. Describe assessment of economic viability of a project.
5. Write a detailed note on "Decision Making".
6. Explain the various elements of product cost with the help of ladder of cost. How is cost controlled?
7. Describe the main features of various financial reports.
8. How are accounts and stores studied in an enterprise? Explain the main scope and objectives.

Chapter

6 Managerial Aspects of Small Business

6.1 INTRODUCTION

It is difficult to differentiate the role of an entrepreneur from that of a manager in a small-scale unit, which is essentially a one-man show. Initially while establishing a venture, one has to function as an entrepreneur but at the operational stage, he will have to perform the role of a manager also. Technical capabilities alone are not adequate to succeed in an industry; other skills also are needed. Effective managerial capability is a key factor in the success of small-scale industry.

6.2. DEFINITION OF MANAGEMENT

Management has been defined as an art of getting things done through people. Management is a distinct process consisting of activities such as planning, organizing, leading and controlling to achieve the desired results. The goals of the enterprise are fulfilled through the use of resources like men, money, materials and machines. Managerial skills are required to exploit these resources and earn profit, face competition and ensure long survival and growth of an enterprise. Throughout the management function, these objectives must be kept in mind.

6.2.1. Principles of Management

1. Division of Work

It promotes efficiency; it ensures work to be executed in limited field.

2. Authority and Responsibility

Authority and responsibility should always go together.

3. Discipline

Poor performance is due to lack of discipline which means obedience, application, energy and respect.

4. Unity of Common

A subordinate should take orders from one supervisor only.

5. Unity of Direction

Each management objective should have one head and one plan.

6. Subordination of Individual Interest to General Interest

Interest of organization is higher than interest of individuals.

7. Remuneration of Personnel

Payment to workers should be fair. There should be a proper method of payment.

8. Centralization

There should be one central power in the organization with power and control of overall work.

9. Order

A place for everyone and everyone in place.

10. Equity

Kindness and justice create loyalty and devotion among employees for the management.

11. Stability

Stability ensures efficiency.

12 Initiatives

Planning before actual execution of work ensures success.

6.2.2. Functions of Management

Management is a function beyond planning, so as to ensure how far the plans are performed and whether these needs to be changed depending upon changing environment and conditions. It reflects flexibility, adoptability and understanding to act and take decisions according to need.

Major tasks of management involve the following:

1. Forecasting

Forecasting is concerned with the estimation of quantities, the sales forecast, cost, finance requirement, capital expenditure, profit and loss, etc. Alternative courses of action should be possible. After selecting the most profitable course of action, planning is started.

2. Planning

Planning is "thinking before doing". Before starting the work, it is decided what is to be produced, how much is to be produced, how to be produced, when to be produced, and who will produce. The planning also aims at elimination of wastage of material, idleness of men, machinery and capital. Policies and objoctives, programme of production and methods of production are decided before hand.

3. Organizing

After completion of planning, the next step is to arrange the material, money and work for actual execution or production. This function consists of material organization and human organization.

4. Directing

The next step is directing the plan into operation. The person must have dynamic leadership for giving direction and guidance to the subordinates. The instructions must be simple, clear, complete and reasonable in writing. The staff must be made conversant with standard operating procedures.

5. Motivating

A capable person without motivation will not give his full output. Motivation is the skill of the manager to encourage a person work and have high morale. Motivation is linked with coordination and cooperation.

6. Co-ordination

A large number of persons may be engaged in producing a product. All the involved persons be co-coordinated, integrated and harmonized to achieve the common objective. The manager or supervisor requires a special skill for coordination.

7. Controlling

All the functions must be performed perfectly alright. The management must control so that actual performance matches with the plans without any defects. The shortcomings must be rectified and these should not be repeated. The following controls are needed: Quantity control, cost control, material control, production control. Some standards are needed for performance evaluation and control.

8. Communication

Lot of information and instructions have to be transmitted to employees, customers, suppliers, shareholders and general public. Management must stress on better communication system to avoid chances of mistruct and fear. Joint consultations and works committees are formed for the exchange of views between employer and employees.

9. Leadership

All managers must have leadership qualities.

10. Decision-making

This is the most important function of the management. The efficiency of the organization, loss and profit are affected by the decisions of the management. The managers have to take decisions very frequently. The entrepreneur when performing the job of a manager requires managerial skills. Major skills required for effective management involve communication skill, inter-personnel skill, problem-solving skill, and a sense of time management. Entrepreneur will have to interact with many people and in different situations with limited time to make decisions, which would demand all these skills.

6.2.3. Necessity of Management

Management is one of the major reasons of failures in small or large industries. Bad management will lead to under-performing the managerial functions of planning, organizing, co-coordinating and controlling the activities. Entrepreneur has to play the role of an all rounder at operating stage.

Management by crisis means last minute decisions and actions. Such management should be avoided. Good management is managing the activities from entrepreneur's viewpoint and also internalising the system in the organizations.

The management may be specified into Financial Management, Marketing Management, Production Management and Personnel Management. However, for a small scale entrepreneur, it is overall management of the above activities rather specializing in one area.

6.2.4. Symptoms of Bad Management

It is very important to know that the common symptoms of a bad management system before it is too late.

1. Haphazard and unplanned activities.
2. Dissatisfaction and lack of motivation on the part of employees.
3. Absence of clear-cut line of action with expected result.
4. Tension and pressures on entrepreneur's mind because of not delegating responsibilities.
5. Frequent breakdown and problems and absence of smooth running of an enterprise.
6. Unrealistic and optimistic view of own performance.

6.3 PLANNING AND PRODUCTION CONTROL

The production planning and control consists of two different and distinct functions:

1. Production planning
2. Production control.

Production planning is involved with the following:

1. Forecasting

It is concerned with estimation of quantities, the sales forecast, cost, finance requirement, capital expenditure, profit and loss, etc. Alternative courses of action should be possible. After selecting the most profitable course of action, planning is started.

2. Planning

Planning is "thinking before doing". Before starting the work, it is decided what is to be produced, how much is to be produced, how to be produced, when to be produced, and who will produce. The planning also aims at elimination of wastage of materials, idleness of men, machinery and capital. Policies and objectives, programme of production and methods of production are decided before hand.

Production control consists of the following functions:

1. Organizing

After completion of planning, the next step is to arrange the material, money and men for actual execution or production. This function consists of material organization and human organization.

2. Directing

The next step is directing the plan into operation. The manager must have dynamic leadership for giving direction and guidance to the subordinates. The instructions must be simple, clear, complete and reasonable in writing. The staff must be made conversant with standard operating procedures.

3. Co-ordination

A large number of persons may be engaged in producing a product. All the involved persons have to be co-ordinated, integrated and harmonized to achieve the objective.

4. Controlling

All the functions must be performed perfectly alright. The actual performance should match with the plans without any defects. The shortcomings must be rectified and these should not be repeated.

The following controls are needed:

Quality control, cost control, material control and production control. Some standards are needed for performance evaluation and control.

6.3.1 Advantages of Effective Production Planning and Control

Effective production planning and control can result in the following achievements:

1. Plant Capacity

All the resources as men, materials and machines are put to maximum use. Effective utilization of plant capacity results in reduced costs and higher returns to the organization.

2. Regular Production

Maximum utilization of resources, ensures regular production. This helps proper delivery of supplies to the customer.

3. Estimation of Resources

The requirement of men and materials are estimated on the basis of orders or sales forecast. Production planning is done to meet sales requirements.

4. Inventory Control

Stocks of raw materials are maintained at proper level to meet production demands. The inventory of finished goods is maintained to ensure regular demand from the market.

5. Coordination

Proper production planning facilitates coordination of activities of various departments. This results in profitable operations in the organisation. Coordination between marketing department and production department is very important. Marketing department gives feedback to the production department regarding quality and quantity requirements of the customer.

6. Minimum Wastages

Proper inventory control and material handling help in minimizing material wastage. Effective control on quality results in minimum rejections.

7. Labour Productivity

Labour efficiency is improved by proper procedure planning and control. Manpower is optimally utilized. Workforce can be given necessary training. Profits are increased. The economic benefits are passed on to the worker in the form of bonus and increase in wages. This motivates the workers resulting in higher labour productivity.

8. Delivery Schedules

There is regular and steady flow of quality products due to effective production planning and control. Delivery schedules are maintained resulting in better customer satisfaction and profitable company operations. The company can face the market competition effectively.

9. Work Environment

Proper production planning and control helps the company to operate at higher efficiency. Workers get improved working conditions, proper working hours, leaves and holidays and better wages and other incentives.

10. Quality Improvement

Proper production planning and control ensures effective and improved inspection and quality control. Quality consciousness of employees is improved.

Effective production planning and control, therefore, helps to improve the efficiency, profitability and health of the organization.

6.3.2. Functions of Production Planning and Control

The functional organization of production planning and control department is given in Figure 6.1.

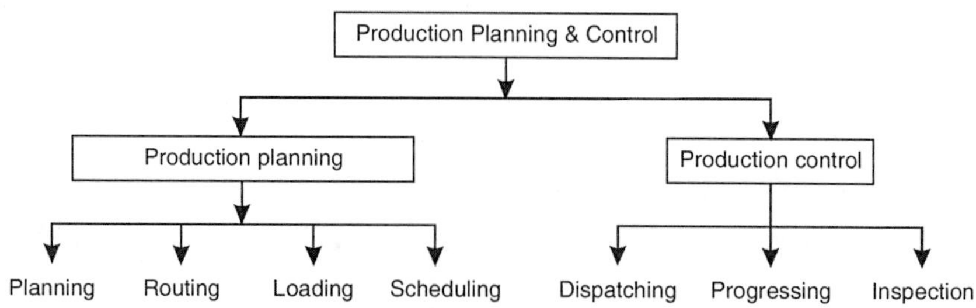

Fig. 6.1. Functions of PPC department.

1. Planning

After finalization of equipments, process and operating parameters, next step is to plan for the requirement of the customer, *i.e.*, quality and delivery schedule of the product.

(*a*) ***Delivery Schedules:***

(*i*) The production planning schedule should be worked out to estimate the delivery schedule. The production capacity of plant and machinery and manufacturing process dictate the delivery schedule.

(*ii*) The production plans should be broken down into weakly, daily and shift-wise production.

(*iii*) Detailed work schedule and production output are needed to work out the raw material and process material requirements.

(*iv*) The work schedule will be the bases for planning of finance and marketing requirements.

(*b*) ***Market Needs:***

(*i*) The data and information on market requirements are important inputs for production planning and control.

(*ii*) Production planning is difficult as dependability of market data is always doubtful.

(*iii*) Multi-product production and job orders complicate the production planning process.

(*iv*) Proper study of market and anticipation of customer's requirements will affect production planning.

(*v*) The size of the enterprise, nature of the product, and requirement of human beings of different specialization will affect the extent and depth of market study.

(*vi*) Detailing of production planning should be decided at the top level.

(*vii*) Production plans, once made have to be monitored and transformed into results.

(*viii*) These plans will take care of finance, material and mankind requirements.

(*c*) ***Inventory Control:***

(*i*) Strict inventory control may not be that important for the smooth functioning of a small-scale unit. However, this control is critical for large enterprises.

(*ii*) Maintenance of required stock levels and identification of sources of supply are important to ensure undisturbed production.

(*iii*) It is dangerous to overstock the inventories.

(*iv*) It will be better to lose an order rather than overstock the inventories.

(*v*) Irregular orders should be avoided in the beginning of operation of a new unit.

The system and practices of a company including quality control, production planning and control, safety of men and equipment, all depend upon the work culture of the company.

2. Routing

Routing is fixing the optimum route of travel or sequential path of flow of all the components of a product from raw material stage to final stage along different departments and machines.

The routing depends upon:

(*i*) Design or shape and size of the component.

(*ii*) Sequence of operations of manufacturing.

(*iii*) Factory and plant layout.

(*iv*) Type of production system.

The problem of routing is simple in case of automated mass production plant. The layout is process layout and the raw material enters from one end and final product is received at the other. The route is decided once in the beginning.

In case of assembly type of industry such as automobiles, cars, scooters, cycles, radio, typewriter, watch, etc., all components should reach the assembly point in time to avoid disruption. The routing is very critical and work-flow sheet is prepared for every component. Skilled and experience persons can prepare this type of routing.

For job orders, after preparation of detailed drawing and planning, sequence of routes for the job is prepared. The production planning and control department must have expert staff.

3. Loading

Assigning specific jobs to men, machines and work centres as per their capacities is called loading.

The following information is required for preparation of loading chart:

1. Time needed for a job.
2. Total work load.
3. Capacity of each worker or grade of worker.
4. Capacity of each machine.
5. Availability of labour.
6. Availability of machines.

Loading Chart

(*i*) A loading chart is prepared to show planned utilization of men and machines available.

(*ii*) A loading chart help to avoid over loading and underloading.

(*iii*) A loading chart helps to avoid bottle necks in the production.

(*iv*) A loading chart helps to achieve maximum utilization of production facilities.

(*v*) The loading chart helps complete of each operation in time.

4. Scheduling

Preparation of time table for production by the planning department is called scheduling. Time table fixes the time and date for each operation as well as sequence of operation. This is somewhat similar to railway time table. A railway map shows the routes followed by the train and time table gives the exact timing of start and reaching at various stations for each train.

There are mainly two types of schedules:

1. ***Master Schedule:*** It is a weekly or monthly breakup of production programme and is prepared as soon as order is received depending upon the availability of men and machines. This schedule is used by the production manager for
   — Advance planning,
   — Check of production rate, and
   — Production efficiency.
2. ***Manufacturing Schedule:*** Shop schedules are prepared from master schedule. It fixes the number of products to be prepared per shift, per day and per week.

5. Dispatching

Releasing of orders and instructions for starting of production as per route sheets and schedule chart is called dispatching. The following functions are started:

1. Required material is issued from the stores to the machines or work station.
2. Work is taken up as per schedule.
3. The different documents are distributed. The important documents are:
   — Machine loading charts.
   — Schedule charts.
   — Route sheets
   — Instruction sheet and forms.
4. Issue of inspection orders showing stage inspection requirements.
5. Ordering of tools, jigs, fixtures and other process material.

Dispatching Documents: Important documents issued with dispatching orders are :

1. **Work order** to different departments to commence particular work.
2. **Time cards** showing time to be taken for each operation. This is issued to every operator for wage payment.

3. **Inspection tickets** are issued to inspection department showing quality of work required, stages of inspection. These tickets are returned with the inspection report accepted or rejected.
4. **Move tickets** are used for movement of material from machine to machine, workstation to workstation or from shop to shop.
5. **Tool and equipment tickets** for issue of tools, gauges, jigs, fixtures and other material required by the shop.

6. Expending or Progressing or Follow-up:

The role of expenditing is to ensure that production is being carried out as per schedule and also help to boost it. This can be carried out as follows:

1. Reporting daily progress of work in each shop in the prescribed proforma.
2. Investigate the causes of deviation from the schedule or delays.
3. Use leadership and motivational skills to boost the morale of the worker.
4. Help to overcome the causes of delay.
   — Production should start in time and material should reach the shops as per requirements.
   — Remove the causes of delay during job progress.
   — All parts should reach the assembly section in actual quantities at proper time.

Causes of Delay: The main reasons of delay can be:

1. Errors in routing, scheduling and dispatching documents.
2. Shortage and delay in material availability.
3. Equipment breakdown.
4. Lack of proper tools, gauges, jigs and fixtures, etc.
5. More rejections.

Flow process chart is used for expediting. This gives graphic representation of operations, transportation, inspection, delays and shortages during production.

7. Inspection

Quality control is designing a system of inspection at various stages of production right from the receipt of raw materials and dispatch of product. Quality depends upon the manufacturing equipment, manufacturing practices, attitude and commitment of the people involved and control system. Quality is the output of men, machinery and material.

Stage inspection help to reduce rejection. Machines and tools should also be inspected. Proper work culture ensures positive attitude and commitment of the people involved.

6.4 QUALITY CONTROL

The quality of a product depends upon:

(*i*) The product quality should be acceptable quality (not the best quality) which satisfies the customer.

(*ii*) The acceptable quality should give sufficient profit.

(*iii*) The product quality should be consistent with market need, price, profit and cost.

(*iv*) The quality depends upon the production system.

(*v*) Quality should be measurable with positive and negative tolerances.

(*vi*) Quality control is not merely checking the final product as per specification.

(*vii*) Quality control is designing a system of inspection at various stages right from receipt of raw materials and dispatch of product.

(*viii*) Quality depends upon the manufacturing equipment, manufacturing practices, attitudes and commitment of people involved and control system.

(*ix*) Quality is the output of man, machinery and material.

Inspection is the process of measuring the quality of a product or service in terms of established standards. Quality control is the prevention of defects originating at design, purchase of raw materials, storage, manufacturing and despatch stage and taking corrective action. Quality assurance is providing necessary confidence to the customer and management that product quality is as per customer's satisfaction with economy. Total Quality Management (TQM) consists of quality planning, quality control and quality improvement.

6.4.1. Inspection

Inspection is the art of comparing materials, products or performance with established standards. The defective items are separated from non-defective ones and thus adequate quality is ensured.

6.4.1.1. Inspection Need

The purpose of inspection may be listed as follows:

1. To ensure adequate quality of products by separating defective items from non-defective items.
2. To inspect and locate defects in raw materials and flows in processes. These may cause problems at the final stage of inspection if not detected.
3. During stage inspection if some part is formed defective, further work should be prevented.
4. The operation of the product should be safe.
5. There should be design check by detecting sources of weakness and trouble in the finished product.
6. The complaints from the customer should be minimized by thorough

inspection of the product. The reputation of the organization is enhanced by proper inspection.

6.4.1.2. Types of Inspection

There are various types of inspection because there are many types of products with different quality requirements.

1. Trial Run

All tools, gauges, jigs and fixtures are inspected. A trial run is made and a single piece is produced. It is inspected to see if it is within tolerance limits.

2. First Piece Inspection

In automatic machines, first 3 or 4 products are inspected. If the products are within limits, the work has started correctly and subsequent products will be satisfactory.

3. Working Inspection

The products are inspected at definite interval during processing to ensure that these are being produced correctly.

4. Sample Inspection

A certain percentage of items are inspected from a lot, which is judged from the result of the sample.

5. Operation Inspection

The work is inspected at the end of a process and before it is sent for another operation or machine or department. The correctness of operation is established.

6. Final Inspection

The final product is inspected before sending to stores.

7. Pilot-piece Inspection

A pilot product is passed through complete manufacturing cycle and every machine and tool are inspected for any defects. The defective tools and machines are adjusted or replaced. The production line is allowed for actual production when a good product starts coming out.

8. Key-operation Inspection

Key-operations are sensitive operations where operators are prone to commit mistakes. Inspection is carried out before and after such operations.

9. Functional Inspection

The complete assembly is ready and inspection is carried out on the assembly to ensure that it will work as required.

10. Endurance Inspection

The assembled product is tested to estimate the time during which the final product will withstand its quality.

11. Patrolling Inspection

The inspection is carried out near the machine while the inspector is keeping an eye on the work during processing.

12. Centralised Inspection

The products are sent to inspection rooms where precision measuring devices are available.

13. Combined Inspection

The modern approach is combined system of floor and centralized inspection method.

6.4.1.3. Stages of Inspection

The inspection is carried out at different locations:

1. Inspection of Incoming Materials

(*i*) The bulk material purchased is economically inspected by resident inspectors at supplies works. Steel, wood, coal, etc., are inspected by resident inspectors.

(*ii*) Specialized equipment and components are inspected by third party inspection. Boilers, nuclear reactors and the components can be entrusted to third party inspection.

(*iii*) Important raw materials and boughtout components can be inspected at the plant before sending to the stores.

2. Stage Inspection

(*i*) In process materials are subject to floor inspection at each machine or work stations.

(*ii*) Strategically located inspection rooms can be used for stage inspection by centralized inspection system.

3. Final Inspection

Most finished goods are inspected immediately before their packing. The final inspection can be carried out by manufacturer's inspector or customer's inspectors or both or by third party inspectors depending upon the type of product and terms of the work order.

Role of Operator

The operator has to contribute to inspection in the following ways:

1. He has to maintain standards of products as per specifications.
2. He has to cooperate with the inspectors.
3. He has to accept the defects of inspection and carry out rectification.
4. He has to minimize the wastages during production.
5. He has to periodically inspect machines and tools to avoid abnormal happening.

Role of Inspector

An inspector has to play the following roles:

1. He must know the inspection standards for raw materials, work-in-process and final products.
2. He must know the properties and degree of finish of products.
3. He must assure that all operations have been performed.
4. He must reject the bad items.
5. He must prepare inspection reports.
6. He must be able to guide to minimize rejection and wastage.
7. He must establish the regulation of the organisation by not compromising on the quality of the products.

6.4.2. Statistical Quality Control

Statistical Quality Control (SQC) is a technique of estimating the quality of the whole from the quality of samples. The laws of chance and statistics are the basis for attaining a state of control in the production process. The concept of statistical control is:

1. Statistical quality control has been a tool used to *control the quality* of end products based on inspection. It is assumed that the defects are normally distributed.
2. In recent years the emphasis has shifted to *preventing defects from occurring* during production cycle. For example, in JIT system, the operators inspect and control quality at all critical work stations.
3. Infections are made to assess the quality of entire population by *examining the random samples.* The purpose is to ensure that the number of defects is below a maximum permitted number.
4. Control charts are used to *analyze the process trend,* to enable appropriate actions to be taken before the process goes out-of-limits set by the quality standards.

6.4.2.1. Control Limits (Process Capacity)

A normal distribution curve is shown in Figure 6.2.

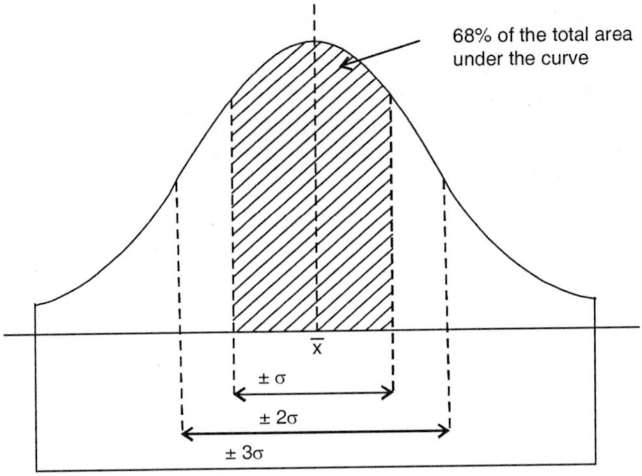

$\overline{X}$ = Arithmetic mean of distribution curve.

σ = Standard deviation.

$\overline{X} \pm \sigma$ represents 68% of the area.

$\overline{X} \pm 2\sigma$ represents 95% of the area.

$\overline{X} \pm 3\sigma$ represents 99.7% of the area.

Fig. 6.2. Normal Distribution curve

The normal curve is symmetrical and bell shaped, 2σ limits occupy 95 per cent of the area of the normal curve and indicate that one is 95 per cent confident that a random observation will fall in this area.

For plotting control charts 3σ limits are selected and are called control limits as shown in Fig. 6.3.

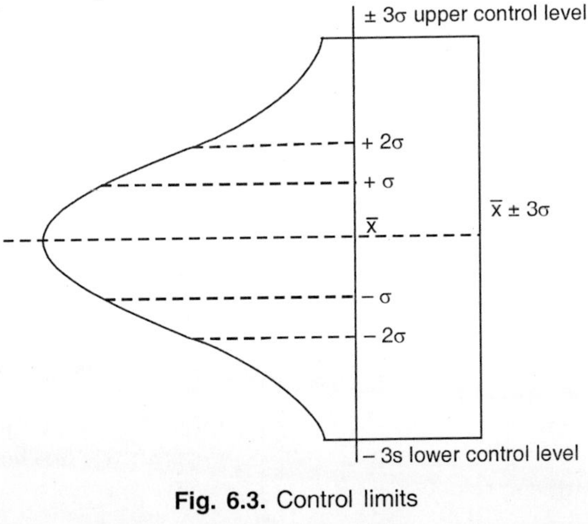

Fig. 6.3. Control limits

The process of statistical quality control with ± 3 σ indicate that one is 99.7 per cent confident that a random observation will fall in this area. This is called process capability.

6.4.3 Total Quality Management

Total quality management can be defined as management philosophy and company practices that aim to harness the human and natural resources of an organization in the most effective way to achieve the objectives of the organisation.

TQM is meeting the requirement of the internal and external customers consistently by continuous improvement in the quality of work of all employees.

6.4.3.1. Concept of TQM

TQM can be conceptualized into the following three processes:

1. **Quality Process:** To understand and satisfy the needs of the customer.
2. **Management Process for Continuous Improvement:** The process comprises continuously evolving policies, objectives and methods to achieve goals, education and training, implementation, checking causes, checking effects, taking appropriate action, and preventing recurrence. The continuous improvement should keep pace with the
   (*i*) Changing requirements
   (*ii*) Competitive environment, and
   (*iii*) Technological advances.
3. **People Process:** It is initiating and maintaining TQM. It is carried out through involvement of all employees on the basis of all the three values, namely, intellectual honesty, self-control and respect for others.

6.4.3.2. Companion of ISO and TQM

ISO is not equal to TQM but is only a part of TQM. Certification for ISO 9000 does not mean that a company has become a world-class company. The differences between ISO 9000 and TQM are presented in Table 6.1.

TABLE 6.1. Differences between ISO and TQM

ISO 9000	TQM
1. Not necessarily customer focussed.	1. Definitely customer focussed.
2. Not integrated with corporate strategy.	2. Integral to company strategy.
3. Technical systems and procedures focussed.	3. Philosophy, concepts, tools and techniques focussed.
4. Employees involvement is not necessary.	4. Emphasis is on employee involvement and improvement.
5. No focus on continuous improvement in ISO 9000. It is a decision.	5. Continuous improvement and TQM is a never-ending journey.
6. Can be departmentally focussed.	6. Organisation wide focus on all departments, functions and levels.
7. Quality department responsible for quality.	7. Everyone is responsible for quality.
8. More likely to preserve the status quo.	8. Involves process and culture change.

(Contd...)

ISO 9000	TQM
9. Three-step maxim of ISO 9000. (*i*) document what you do (*ii*) do only what you document (*iii*) demonstrate that you have done it, by documentary proof	9. Customer satisfaction and economic cost are TQMs two distinguish features.
10. ISO 9000 are technical and physical systems and standards.	10. TQM is a philosophy where the approch is behavioural and human.

6.4.3.3. Advantages of TQM

In aggregate, TQM offers the following advantages:

1. Increased efficiency in processes.
2. More time for innovation and creativity.
3. Higher morale of employees.
4. Improvement in the quality of products and services.
5. Increased customer satisfaction.
6. Larger market share.
7. Higher productivity.
8. Higher profits.

6.4.3.4. TQM Process

The evolution and development of the quality concept over the period is shown below:

Quality Imperatives	Period
1. Operator Inspection	1870 – 1890
2. Foreman Inspection	1890 – 1920
3. Inspection Department	1920 – 1940
4. Statistical Quality Control	1940 – 1960
5. Quality Assurance Department	1960 – 1990
6. Total Quality Management	1990 – onwards

Of late, the concept of quality has expanded to TQM.

TQM involves the following processes shown in Fig. 6.4.

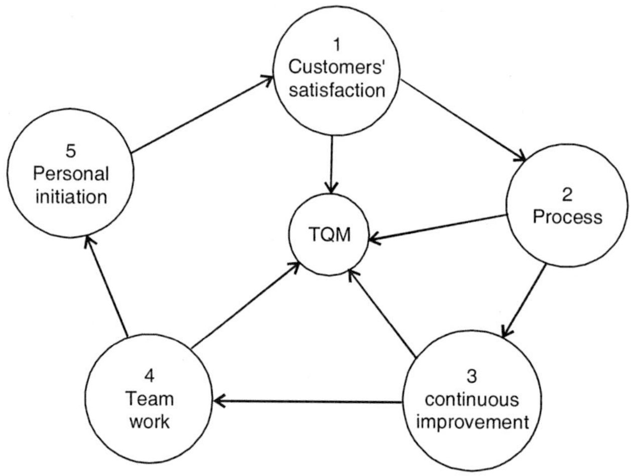

1. Customer satisfaction — understanding their needs.
2. Processes — to be defined.
3. Continuous improvement — reduction in cycle time.
4. Team work — effective.
5. Permanent Initiative — encouragement.

Fig. 6.4. TQM Process.

6.5 MARKETING

A market survey must be conducted to ascertain the true demand and supply position. Information on other entrepreneurs planning to enter the same field must be collected.

Once it is established that there is good possibility of the unit being able to establish itself, a pre-production market development programme should be undertaken. This is to ensure that the enterprise does not face the problem of want of sales after commencement of production. The role of marketing starts even before taking a decision to set up a manufacturing unit.

6.5.1. Marketing Activities

Marketing comprises of the following activities:

1. Research
2. Planning
3. Branding
4. Pricing
5. Distribution
6. Selling
7. Packaging
8. Merchandising
9. Warehousing
10. After-sales service
11. Sales promotion
12. Advertising
13. Credit control

1. Research

Market research is a continuous activity. It starts with a market survey to establish demand and supply position for the product. This is required to prepare feasibility report for the establishment of manufacturing units. After feasibility is established, research is required to identify unsatisfied consumer needs. The product planning and manufacturing should be designed to meet the gap of unfulfilled needs. Consumer satisfaction should be the main criteria for evaluation of policies and action.

When the enterprise is under production market research has to continue to improve the product, packaging, distribution, etc.

2. Planning

During establishment of the enterprise, planning, market development, packaging development, distribution network development, warehousing, short-term and long-term sales planning are to be carried out.

When the unit is under production, sales planning is also included. This will comprise of a sales forecast, a plan for territorial coverage, a programme to achieve, the projected sales and a sales promotion plan.

3. Branding

Branding is essential for consumer products, consumer durables as well as industrial products. Same product with widely diverse specification is available from different manufacturers.

Branding helps the manufacturer, retailer and consumer in the following manner:

(*i*) The consumer can effectively and easily identify the product of a particular manufacture.

(*ii*) The manufacture is benefited as the consumer demands specific brand.

(*iii*) Manufacturer can have better control over the sales outlets and product price.

(*iv*) The retailer uses a popular brand to attract customers and he can sell other products also.

When selecting a brand name, remember this word RIPPS :

R – Remember. Name of brand should be easy to remember.

I – Image. The name should create an image of the product in the minds of consumers.

P – Pronounce. It should be easy to pronounce the name.

P – Protectable. The name should not be easily copied without infringing patents

S – Short. The name should be short.

4. Pricing

The pricing is usually associated with:

(*i*) Value
(*ii*) Status
(*iii*) Quality
(*iv*) Durability

Consumer is more sensitive to price when the purchase frequency is high. Also when the quality of competing product is more or less the same. Examples: Hardware, matches, gums, stationery, chemicals, petroleum products, etc.

The price is immaterial when there is quality variations in brands.

Methods of Pricing

(*i*) Pricing in line with similar products existing in the market.
(*ii*) Pricing higher than the price of leader in the market.
(*iii*) Pricing slightly lower than the price level in the market.
(*iv*) Pricing on cost plus basis.
(*v*) Pricing on the basis of profit yield at different sales levels.

Factors of Pricing

(*i*) Price should be economically attractive to customer.
(*ii*) Different price levels will result in different sales volume.
(*iii*) Reaction of the competitors.

5. Distribution

The distribution network should make available goods conveniently to largest number of consumers in required quantities, where and when needed: This can be achieved through.

(*i*) Agents
(*ii*) Stockists
(*iii*) Whole-sales
(*iv*) Distribution
(*v*) Manufacturer's shops
(*vi*) Retailers
(*vii*) Street vendors
(*viii*) Mail orders

The selection of sales outlets will depend on nature of the product, number of consumers and the financial resources.

6. Selling

Converting goods into cash is called sales. No sale is complete till the goods have been delivered and their value recovered. The most effective method of selling is through own trained staff. Own sales staff knows the plus points of the product better than any agent, distributor, stockist or dealer. Personal selling is possible only if:

(*i*) The product is to be sold to a small limited number of customers, and
(*ii*) The demand for the product is large and the sales are restricted to an overall area in the vicinity of the manufacturing unit.

Compensation to Own Salesmen

It showed include the following:

(*i*) Salary

(*ii*) Salary-cum-commission
(*iii*) Commission-cum-retainership
(*iv*) Commission

Motivation Methods

(*i*) Commission
(*ii*) Merit awards
(iii) Salesman's meetings
(iv) Contests

Maintenance of Effective Control

Different methods for maintaining effective control on own sales force:

(*i*) Sales reports
(*ii*) Personal meetings
(*iii*) Field visits
(*iv*) Fixing of sales quotas
(*v*) Evaluation on the basis of compensation earned
(*vi*) Territorial development.

The sales staff of outside agencies should be given suitable product training. Also sufficient sales aids should be provided to make them more effective. The outside agencies can be motivated through:

(*i*) Volume discounts in addition to normal discounts
(*ii*) Contents
(*iii*) Dealer meetings

7. Packaging

The packaging is used:

(*i*) To hold the product.
(*ii*) To protect the product.
(*iii*) To attract customers.

The type of packaging should meet the above requirements. However, the selection of design of packaging should also take into account the cost factor.

8. Merchandising

Merchandising is the art of displaying products at a point of sale in order to:

(*i*) Catch the attention of the customer.
(*ii*) Highlight the plus points of the product.
(*iii*) Lure customer into buying the product.

This is used mainly for consumer goods and durables.

Methods of Merchandising

(*i*) Display windows.
(*ii*) Display on the shelves and sales counters.
(*iii*) Hoardings and bill boards outside the sales outlets.

9. Warehousing

Warehousing is to store goods for safety against theft, weather, rodents and insects, etc. The important reasons of warehousing are:

(*i*) Storage of products outside octroi limits results in financing savings.
(*ii*) Warehousing of products so that inter-state selling is possible, the lower inter-state sales tax rate results in financial savings.
(*iii*) Warehousing of goods in the proximity of market can save in delivery period resulting in increased sales.
(*iv*) Warehousing of goods near consumers at lower cost, if possible, can result in financial savings.
(*v*) Operating excise bonded warehouses can result in financial savings due to saving of interest on the duty value for duration of storage.

The above factors should be considered for finalization of warehousing arrangements.

10. After Sales Service

This is a very important activity usually ignored by many organizations. Dissatisfaction of the customer can very seriously affect future sales. Adverse reports spread very fast. Prompt action in solving customer's complaints can win customer praise resulting in increased sales.

11. Sales Promotion

Sales promotion is required to create temporary spurt in sales to raise finances in short time. The following methods are used for sales promotion.

(*i*) Trying up sales of the product with fast moving product.
(*ii*) Special discount for a specific time period.
(*iii*) Tying up sales of the product with a fast moving product of other manufactures at a special discount.
(*iv*) Contest with attractive prizes for winners with entry fee as cash memo, wrappers etc. of the product.
(*v*) A raffle for lucky members with attractive prizes.

12. Advertising

Advertising in dailies and magazines and on radio and television is most popular but expensive. There are, however, less expensive avenues also. The different media are:

(*i*) Newspapers and magazines
(*ii*) Radio
(*iii*) Television
(*iv*) Hoardings and neon signs
(*v*) Bus and train panels
(*vi*) Bill boards
(*vii*) Cinema slides and short film advertisements
(*viii*) Stalls at exhibitions and fairs
(*ix*) Direct mailing of literature
(*x*) Demonstration

A certain provision should be made in the annual sales budget for advertising. The available funds must be spent in most effective manner.

13. Credit Policy

Credit policy is very critical for the survival of a small-scale unit.

(*i*) The small-scale units have limited finance. The money tied up in the credit results in reduced available money for purchase of raw materials and other inputs.

(*ii*) The turnover is reduced due to less materials purchased.

(*iii*) The banks are very stringent in granting facilities to small-scale units.

(*iv*) It is better to give 2% to 2.5% cash discount if cash sales in preference to extending credit of 30 days or more.

(*v*) Even after agreeing to 30 days credit, the customer usually stretches the priced resulting in non-availability of resources to the unit.

(*vi*) Do not extend credit without security even when dealing with distributors, stockists or dealers.

(*vii*) The customer should have the ability and desire to pay.

(*viii*) Credit should be extended to customers with long business relations.

(*ix*) Fix the value and time limit to each debtor when extending credit. Don't allow the debtor to stretch the agreed term. A single relaxation can become a permanent feature.

(*x*) Regular reminders before and on the due date must be the policy.

(*xi*) Regular statements of accounts must be posted to the party to avoide any differences later on.

14. Important Tips

(*i*) Even a small amount earned when multiplied by the number of items sold can result in big amount. Therefore, a rupee saved is a rupee earned.

(*ii*) The annual budget must be broken into monthly budget. The actual performance in each month must be evaluated against the budget. Corrective steps must be taken if variance is more than five per cent.

(*iii*) Quality consciousness and building up reputation of the brand will pay dividends in the long run.

(*iv*) Clean and honest dealing with both suppliers and customers will build up reputation.

6.6 INDUSTRIAL RELATIONS

Industrial relation is the management function dealing with manpower which includes all workers, operators and managers. Managing the men is the most difficult management function. It's much easier to manage machines and materials. Human feelings and emotions are involved in human relations that require proper attention. The management should know the behaviour of persons in different situations.

Cordial and peaceful industrial relations are highly essential for increasing production and economic growth of the industry. Poor human relations are responsible for troubles like strikes, protests, agitation, frustrations and labour turn-over.

Prof. John Mee has tabulated human relation problems and their possible reasons. These are reproduced in Table 6.2.

TABLE 6.2. Industrial Relation Problem

S. No.	Problem	Possible Reasons	Possible Action
1.	Absenteeism	Habit, lack of interest, Trouble at home, dissatisfaction with job, Transportation trouble.	Investigate, have personal talk, create interest in work, help to get transport.
2.	Complaints about working conditions	Jealousy, favouritism, actual poor conditions.	Better instructions, check supervision, transfer to other work.
3.	Dirty work and Dirty work-place.	Lack of interest, Carelessness, toleration by management.	Personal talk, explain safety hazards, check supervisor's tolerance, allow cleaning time.
4.	Disloyalty	Discontent, favouristism, lack of recognition.	Investigate the reason, Setup an objective merit rating scheme.
5.	Complainer spreading discontent to others	Work conditions, poor wages, lack of recognition, poor leadership.	Transfer, investigate the wage level, check if work is recognised.
6.	Excessive talking	Lack of interest, jealousy, favouritism, personal issue.	Redefine tasks, make employee realise that talking disturbs others.
7.	Insubordination	Ignorance of facts, dissatisfaction.	Investigate the reason, Transfer, if the reason is personal, try talking to both the parties.
8.	Misinterpreting management information	Indifference, desire to turn out more work, device improperly designed.	Better communication, Give facts is simple language.
9.	Not using safety Aids	Clownishness, desire to be important to group, lacks discipline, not enough work.	Show safety movies, talk about personal danger, check safety aid design, Penalise.
10.	Plays on job	Clownishness, desire to be important to group, lacks discipline, not enough work.	Show safety hazards, Get group action, Check the assigned work, Allow break time.
11.	Quality of work poor	Lack of knowledge, laziness, desire to exceed price rate.	Show the importance of the job to final product, proper motivation, Emphasise need for quality, Penlise for poor work, order rework.
12.	Rumour spreader	Jealousy, Curiosity desire to seem importance.	Investigate, recognise work, personal talk.
13.	Leaving office before time	Habit, transport difficulty, lack of washroom facility, desire to get out.	Insist in quality work, provide adequate bathroom facilities, check leaving time.

6.6.1. Industrial Disputes

Definition and Introduction

- An *Industrial Dispute* means any dispute or difference between employers and employees or employers and workmen or between workmen and workmen which is connected with employment or non-employment or terms of employment or conditions of labour, of any person.
- Every human being (say a worker) has certain needs, *e.g.*, economic needs, social needs and needs for security. When these needs do not get satisfied, there arises a conflict between labour and capital. A conflict means a struggle or clash between the interests of the employer and the workers. For example, in order to compete in the market, the employer would like to reduce the price of his product and for doing so he will tend to reduce the wages of the workers; the workers would not agree to it and the result will be an *industrial conflict* between the employer and the workers.
- When an industrial conflict (which otherwise is general in nature) acquires a concrete and specific display or revelation, it becomes an *Industrial Dispute.*
- A conflict takes the shape of Industrial Dispute as soon as the issues of controversy are submitted to the employer for negotiations.
- An industrial dispute may be looked upon as a controversy or disagreement between employer (or management) and the workmen on issues such as Wages and other benefits and work hours and working conditions, etc.
- Industrial disputes cause losses to, workers, management and nation as a whole.
  (*i*) Workers lose their wages.
  (*ii*) Management loses its profit.
  (*iii*) Public suffers due to shortage of goods in the market.
  (*iv*) Nation suffers due to loss of production.

Causes of Industrial Disputes

Some of the common causes of Industrial Disputes have been listed below:

(*i*) Psychological causes
- Difficulty in adjusting with each other (*i.e.*, employer and worker).
- Clash of personalities.
- Authoritarian Leadership (administration).
- Demand for self-respect and recognition by workers.
- Strict discipline.

(*ii*) Institutional causes
- Non-recognition of the labour union by the management.
- Matters of collective bargaining.
- Unfair conditions and practices.

– Pressing workers, not to become members of union, etc.

(iii) Economic causes

(*a*) Terms and conditions of employment — More hours of work.

– Working in night shifts.

– Promotion, layoff, retrenchment, dismissal, etc.

(*b*) Working conditions

– Environmental conditions such as too hot, too cold, noisy dirty, messy, etc.

– Improper plant and workstation layout.

– Old and trouble giving machines.

– Frequent changes in products, etc.

(*c*) Wages and other benefits

– Inadequate wages.

– Undesired deductions from wages.

– Poor fringe benefits.

– No bonus or other incentives, etc.

(iv) **Denial of legal and other rights of workers**

– Not proceeding as per labour laws and regulations, standing orders etc.

– Violation of already made mutual agreements (*i.e.*, between employer and workers).

6.6.2. Agitation, Strikes, Lockouts, Picketing and Gherao

1. Agitation

In the event of any grievances towards employer, at times the workers feel so anxious, nervous or upset that they cannot think calmly. This is the state of agitation.

2. Strikes

Introduction. Whenever workers feel any grievance and if the same is not removed by the management, the workers unite together to fight it out and this causes industrial unrest, conflict and dispute. The industrial disputes generally results in **Strikes, Lockouts, Picketing or Gherao.**

– The word *strike* is an innovation of the early 19th century but the phrase to *strike work was* used in the eighteenth century.

– Strike is the ultimate weapon of the trade union by which it can threat the employer.

– Strike implies that

(*i*) there shall be cessation of work by a body of workmen; and

(*ii*) workmen should be acting in concert in order to enforce a demand against the employer during an industrial dispute.

– Success of strike depends on the ability of the workmen to stop the employee from continuing to operate.

3. Lockout

Introduction and definition. It is the weapon of employer to pressurize workers to come down in their demands. A lockout means 'closing the place of employment or suspension of work or the refusal by an employer to employ any number of persons employed by him.

4. Picketing

Introduction and Definition:

- Picketing is almost a standard practice when a union strikes against an employer.
- some workers are placed at the factory gate to discourage others from entering the work premises.
- pressure on the employer increases when employees refuse to cross picket line to deliver or pickup goods from the struck employer.
- in picketing workers may parade with banners to inform public about the dispute with the employer and to enlist popular support for the workers or unions.
- picketing may be designed to interfere with business and thus to pressurize the employer to comply with union demands.
- picketing by one group of employees often stops other employees from working in the same factory.
- in doing so, the pickets sometimes insult and even block physically the path of those employees intending to enter the factory.

5. Gherao

- Like strike and picketing, Gherao is also a method to pressurize employer to fulfil union demand.
- In gherao workers force the employer or managers to remain confined in their offices for hours or even days. The employer sometimes is forced to remain without water and food; he is not allowed to go out even for natural calls.
- Workers encircle the office of the employer, close all the exits and sit around in batches.
- In contrast to strikes which put economic pressure on the employer, a gherao involves physical coercion, *i.e.*, it tends to inflict physical duress on the employer or manager.
- Thus a gherao besides endangering industrial harmony, creates problems of law and order also.
- Actually gherao is a primitive method which used to be employed in England by early trade unions.

6.6.3. Settlement of Industrial Disputes

Guidelines are provided in the Industrial Dispute Act, 1947 for handling and settlement of Industrial Disputes.

Industrial Dispute Act, 1947

The whole history of labour struggle indicates the continuous demand for fair return to labour. Industrial disputes are mainly the result of dissatisfaction amongst the labour as regards their existing labour conditions. With the above in mind, the Industrial Dispute Act came into force in 1947 and aims at settling the industrial disputes on a new pattern known under the Act as adjudication machinery.

- This act aims at making industrial peace through voluntary negotiations and compulsory adjudication.

The act makes provision for settlement of industrial disputes between employees (workers) and employer.

Important Aspects of the Act

(*i*) An industrial dispute may be referred to an Industrial tribunal.

(*ii*) An award shall be binding on both the disputing parties for a period not exceeding one year.

(*iii*) Strikes and lockouts are prohibited during the pendency of,
- conciliation and adjudication proceedings;
- settlement reached in the course of conciliation proceedings,
- awards of industrial tribunals declared binding by the appropriate government.

(*iv*) In public interest or emergency, appropriate government can declare the following industries to be a public utility service for a maximum period of 6 months:
- Transport by land, air, and water
- Foodstuff
- Coal
- Cotton textile, and
- Iron and Steel.

6.7 SALES PROMOTION

1. Customers Needs

A customer does not buy a product but an idea or benefit. He buys a CASHED

C – Comfort
A – Approval
S – Safety
H – Health
E – Economy
D – Desire

2. Unique Selling Proposition

The product must have uniqueness. The strong selling point of the product is called Unique Selling Proposition (USP). The USP of the product should be:
- Easy to understand
- Customer oriented
- Easy to remember

3. Selling Philosophy

When a customer buys a product, he wants some benefit:

(*i*) Convince the customer that the product will satisfy his need the best.

(*ii*) Give a good demonstration of the product. Convince him that delay in buying will cost him.

(*iii*) Give warm reception to the customer.

(*iv*) Attract maximum attention of the customer.

(*v*) Create interest in your product.

(*vi*) Create desire to buy your product.

4. Handling Competition

Welcome competition and follow the steps:

(*i*) Know the positive and negative points of the competitor.

(*ii*) Do not be afraid of competition.

(*iii*) Do not be over-confident.

(*iv*) Express that the product of the competitor is good but your product is better.

(*v*) Lowering of price is not the only solution to handle competition. Think of other factors also.

(*vi*) Create an image of your product and try to sell image not the product.

5. Good Salesman

A professional salesman should

(*i*) Have ability to sell

(*ii*) Be loyal

(*iii*) Take interest in selling

(*iv*) Be skilled in his work

(*v*) Have thorough knowledge

(*vi*) Continue learning and updated

(*vii*) Be hardworking

(*viii*) Understand the value of self-respect

(*ix*) Give good service to customers

(*x*) Be creative

6.8 MATERIAL PURCHASE

Material management consists of process of integrating the activities involved in acquisition and utilization of materials required in the production of final product. Purchasing is the first phase of material management. Purchasing is responsible for getting the right material, from right place, in right quantity, at right time and at right price.

Purchase Procedure

The purchase procedure consists of the following stages:

1. Receipt and analysis of requirements from different departments and processing the requisitions.
2. Selection and location of potential suppliers.
3. Request, receipt and analysis of quotations.

4. Placing of orders.
5. Expediting and following of purchase orders.
6. Verification of vendor's invoices for payments after processing discrepancies and rejection.
7. Closing of completed orders and maintenance of records.

Methods of Purchasing

The following are some popular methods of purchasing:

1. Purchasing according to requirements—Hand to mouth purchasing.
2. Purchasing for some specified future period.
3. Market purchasing.
4. Rate contract purchasing.
5. Central purchase organization.
6. Rate contract through D.G.S. and D.

1. Hand to Mouth Purchase

In a job order system, material requirement is worked out after receipts of order. No inventory is kept. This system has the advantage of less capital requirements.

2. Purchasing for Specified Future Period

There are standard items which are consumed regularly. The volume of consumption is low and price fluctuation is not much. These items are purchased and are available as soon as some order is received.

3. Market Purchasing

Market purchasing helps to take advantage of price fluctuations. The purchasing is related to price fluctuations and not production needs. This method has the shortcoming that there can be loss to the organization if price fluctuation assessment is wrong.

4. Rate Contract Purchasing

The organization enters into agreement with various vendors to supply goods in future periodically on rate contract basis. Although prices are negotiated during lean period but there is a provision for price review or cancellation of contract with sufficient notice period.

This method is used for purchase of large quantities of basic raw materials, like coal, pig iron, paints, cables, etc.

5. Central Purchase Organization

Large government and private organizations have central purchase system. The requirements from various departments, shops and sections are collected. Central stores keep supplying the material to sectional stores.

The central purchase can be carried out at cheap rates and better bargain.

6. Rate Contract through D.G.S. and D

The various government departments and other firms can place orders for the supply of requisite materials at the contract rates approved by D.G.S .and D. These suppliers cannot supply materials to other purchasers at lesser rates during the contract year.

6.8.1 Purchase Documents

There are certain very important documents used by purchase department.

1. Purchase Requisition: A purchase requisition is prepared in duplicate by the department requiring material. It is issued to purchase manager for necessary action. A sample of purchase requisition is shown in Fig. 6.5.

Requisition No. ....

Date ....

Work order No. ....

Department ....

The Purchase Manager,

Please order the following materials.

S. No.	Description of material	Quantity	Delivery period	Estimated price	Suggested vendors

Prepared by.................... Checked by.................... Approved by....................

Fig. 6.5. Purchase Requisition Format

2. Performa for Quotation

Request for quotations is made on prescribed performa as per for Fig. 6.6 to all the selected and approved vendors. This is accompanied with purchase conditions

S. No.	Description of material	Quantity required	Price	Discount	Remarks

(Purchase Officer)

Fig. 6.6. Performa for Quotation

3. Comparative Statement

A comparative statement as shown in Fig. 6.7 is prepared on the basis of quotations received. These are analyzed for rates and other terms and conditions of supply.

Department.............................. Dated..................

Enquiry Letter No.................. Dated..................

Date of opening of quotations ....

Comparative statement prepared by........................ Checked by........................

S. No.	Material Specificates	Qty.	Vendor 1	Vendor 2	Vendor 3
				Rates	
1.					
2.					
3.					
4.					
5.					
6.					
7.					
8.					
9.					
10.					
				Terms and Conditions	

Terms and Conditions:

(1) Data and place of delivery.

(2) Terms of payments.

(3) Taxes and excise duty.

(4) Packing and forwarding charges.

Fig. 6.7. Comparative Statement

4. Purchase Order

After selecting the right vendor, a purchase order is prepared as per Fig. 6.8.

Purchase Order No. ....

Date ....

Requisition No. ....

M/s ....

....

....

Refer your quotation No. ...., dated .... and subsequent discussions on ...., we are happy to place the following order on the terms and conditions mentioned below:

S. No.	Material specifications	Qty.	Rate	Amount	Date of Delivery

Terms and Conditions

1. **(Purchase Officer)**
2.
3.

Fig. 6.8. Purchase order

6.9 ADVERTISEMENT

Advertisements provide information to the prospective buyers about an enterprize, its products, product qualities, place of availability of product. It is complimentary to personnel selling.

6.9.1 Functions of Advertising

Advertising is beneficial to manufacturers, traders, consumers and the society. Advertising performs the following functions :

1. Promotion of sales of good and services.
2. Introduction of new products in the market.
3. Create benefit of economy-of-scale as prices can be lowered by increasing production.

4. Build reputation of an enterprize.
5. Educate consumer about new products and their uses.
6. Generates employment.
7. Creates revenue for media.

6.9.2 Advantages to Entrepreneur

Advertising informs, educates and persuades a customer to buy. It has following benefits to an entrepreneur.

1. An enterprize can introduce itself and its products to the public through advertising.
2. It is an effective communication system to develop new taste and stimulate to the public to purchase.
3. It helps to create new markets and attract new customers.
4. It helps to create steady demand of a product.
5. It helps to remain in the market and face competition.
6. It helps to create goodwill by increasing good quality of a product.
7. It helps to boost morale of sales persons and other employees.
8. It helps to increase demand and lower the preduct price by mass manufacture and lower distribution cost.

6.9.3 Advantages to Customers

Advertisement also gives following advantages to customers.

1. It informs the customer about various products and helps him for proper selection.
2. It avoids exploitation of customer by sellers.
3. It educates the customer about new products and their various uses.
4. It adds to the amount of satisfaction of the customer.
5. It induces manufacturers to improve quality of their product.

6.9.4 Requirements of a Good Advertisement

Advertisement has become a specialised job and should fulfill the following requirements.

1. It must attract the attention of the people.
2. It must make the people to remember the message.
3. It should educate the people about the product.
4. It must have convincing force.
5. It should create a need of the product in the minds of the people.
6. It should persuade the people to buy the product.
7. It should suggest to the people something about the product.

6.9.5 Media for Advertisement

Advertising in dailies magazines and on radio and television is most popular but expensive. There are however, less expensive avenues also. The different media are:

1. Newspapers and magazines.
2. Radio.
3. Television.
4. Hoardings and neon sign.
5. Bus and Train panels.
6. Bill boards.
7. Cinema slides and short film advertisements.
8. Stalls at exhibitions and fairs.
9. Direct mailing of literature.
10. Demonstration.

A certain provision should be made in the annual sales budget for advertising. The available funds must be spent in most effective manner.

6.10 WAGES AND INCENTIVES

Compensation is the payment made by the employer to the employees for doing work. In addition to wages, the compensation may include other benefits and services provided to employees. The monetary compensastion is the form of wages plays a more powerful role than non-monetary benefits.

6.10.1 Wages

The salary or wages depend upon the following factors :

1. Demand and supply of labour.
2. Bargaining power of employees or trade unions.
3. Cost of living.
4. Employers capacity to pay.
5. Market rate.
6. Job requirement.
7. Government regulation.
8. Management attitude.
9. Social needs.

Wages are paid to the workers, which include basic wages and other allowances linked with dearness allowance.

1. Minimum Wages

These are paid to workers for base sustenance of life, preservation of efficiency and must include measures of education, medical requirements and amenities.

2. Living Wages

These should enable the employee to provide for food, clothing and shelter for himself and his family living comfort, education for children, protection against ill-health essential social needs, insurance against misfortune and old age.

3. Fair Wages

These are the wages between the minimum wages and living wages. Fair wage is the wage paid to the workers performing equal work with equal difficulties and risk, demanding equal skill and unpleasantness. It is also the rate of wages prevailing for the skill and qualification of the workers.

6.10.2 Incentives

Incentives are used for motivations. These are of two types:

(*i*) Financial Incentives
(*ii*) Non-financial Incentives

1. Financial Incentives

Money can be a very powerful motivation for persons operating at the lower level as it can satisfy the physiological needs effectively. Money can be given in the form of pay, promotion and rewards.

Adequate pay compensation programme will attract and retain key people of superior calibre in the organisations. Such a programme will motivate and stimulate a person to improve his performance. Monetary compensation can be used to reward significant achievements of employees. In addition, fringe benefits including bonus also motivate an employee.

2. Non-financial Incentives

Money can satisfy only physiological needs of a person. The safety and security needs can be satisfied by psychological climate or environment and work culture of the organisation.

The feeling of achievement can motivate a person. Such achievement should be followed by recognition as follows:

(*i*) Providing responsibility through job enlargement
(*ii*) Providing participation
(*iii*) Creating sense of achievement
(*iv*) Providing recognition for accomplishment
(*v*) Offering inducement of promotion and growth as a result of effective performance

6.10.3. Promotion as Motivator

Promotion policy should generate continued motivation to employees.

Promotion should give assignment of a position of greater responsibility and authority.

Promotion should ensure ascending in the management hierarchy.

Promotion should provide increase in pay or status or both.

Promotion should be based on seniority as well as merit.

The *promotion systems* can be

1. Promotion by seniority
2. Promotion by merit
3. Promotion through reports

6.10.4. Job Satisfaction as Motivator

Production increases with job satisfaction of workers and industrial disputes are reduced. The following areas help in increase of job satisfaction.

1. Incentives and profit sharing schemes
2. Promotion and training programmes
3. Reasonable collective bargaining policy
4. Hospital benefits and group insurance
5. Job security
6. Safety and health measures
7. Suggestion plan: Inviting suggestions from workers
8. Housing facilities
9. Counselling for personnal problems
10. Sports and social activities
11. Job encrichment

6.11. INVENTORY CONTROL

Inventories are stocks of the product a company is manufacturing for sale and components that make up the product. Inventories form a link between the production and sale of the product. There are three types of inventories.

(*i*) *Raw Materials*

Goods which have been purchased and stored for future production but have not yet been committed to production.

(*ii*) *Work-in-progress.*

These are semi-manufactured products and have been committed to production but the finished goods have not yet been produced.

(*iii*) *Finished Products*

These are final products ready for sale.

The size of the above types of inventories to be maintained depends upon type of unit and the product.

6.11.1. Need for Inventory Management

The inventory management is required:

(*i*) To make adequate inventories available for production and sale as and when required.

(*ii*) To minimize costs and volume of investment in inventories consistant with above.

There are three major motives for holding inventories.

1. Transaction Motive

An enterprise maintains inventories to avoid bottlenecks in production and sales. The production should not be interrupted for want of raw materials and sales should not be affected on account of non-availability of finished goods.

2. Precautionary Motive

Inventories are also held as a cushion against unpredicted spurt in demand for finished goods or slump in the supply of raw materials.

3. Speculative Motive

Inventories may also be need to take advantage of price fluctuations in the prices of raw materials and finished goods.

6.11.2. Advantage of Holding Inventories

1. Avoiding sales losses on account of non-supply of goods at times demanded by the customers.
2. Ordering costs can be reduced by planning large orders rather than several small orders.
3. Achieving efficient production run and protection against shortage of raw materials, which may interrupt production operation.

However, there are costs of holding inventories:

1. Material costs include ordering cost and cost of material and components.
2. Carrying costs include insurance charges for covering risks, rent for storage space, wages, wastages, deterioration, theft, pilferages, etc.
3. Opportunitiy costs as the money blocked in inventories could be invested elsewhere to earn profit.

Therefore, there is a need for inventory management, *i.e.*, to decide the optimum volume of inventories during the period.

6.11.3. Models of Inventory Management

There are three models in use:

1. Economic Ordering Quantity (EOQ) Model.
2. ABC Analysis.
3. Inventory Turnover Ratio.

1. Economic Ordering Quantity (EOQ) Model

This is based on Baumol's cash management model. EOQ is decided on the basis of "cost of acquiring inventories" and "cost of holding inventories." There are basically three approaches.

(*i*) The Order-Formula Approach:

The EOQ can be calculated

$$Q = \sqrt{\frac{2U \times P}{S}}$$

where Q = Economic Ordering Quantity (EOQ)

U = Quantity purchased in a year or month
P = Cost of placing an order
S = Carrying cost

Example: Calculate EOQ for a unit with the following date:

Annual requirement = 800 unit
Ordering cost = Rs. 50/order
Carrying cost = Rs. 2/unit
Purchase cost = Rs. 100/ unit

$$\text{EOQ} = \sqrt{\frac{2 \times 800 \times 50}{2}} = 200 \text{ units}$$

(ii) Trial and Error Approach

A number of alternatives may be used to determine the volume of inventory and work out EOQ by the order-formula approach. The alternative which gives minimum cost can be selected.

(iii) The Graphic Approach

The ordering cost, the carrying cost and total cost are plotted in figure 6.9. EOQ takes place at the point where the total cost is minimum.

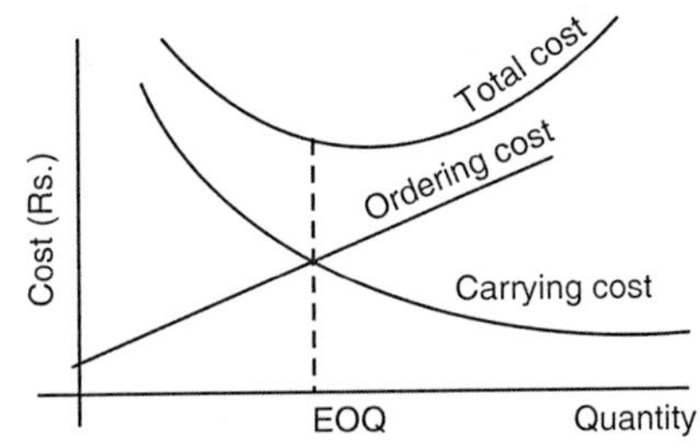

Fig. 6.9. EOQ

2. ABC Analysis (Selective Inventory Control)

Selective approach can be used for various types of inventories. The items with highest value are classified as "A items". The items with relatively low value as "B items" and the item of least value are classified as "C items" Stringent control is imposed on " A items" maintaining bare minimum necessary level of inventories of these items. "B item" will be kept under reasonable control. 'C items" would be under simple control.

3. Inventory Turnover Ratio

$$\text{Inventory Turnover Ratio} = \frac{\text{Cost of goods consumed or sold in a year}}{\text{Average inventory during year}}$$

A comparison of current year's inventory ratio with those of previous years will tell us about Fast Moving Items, Slow moving items and Obsolete items.

Adequate inventory of fast moving items ensure smooth production. Slow moving items may be maintained at the minimum level. Obsolete items should be disposed off as early as possible.

Question Bank

1. What is management ? What are the main principles of management ? Discuss the important functions of management.
2. What are the function of production planning and control ?

3. What is the need for inventory control ? Discuss various models of inventory management.
4. How is marketing different from selling ? Explain various marketing techniques employed.
5. Explain the following:
   (a) Statistical Quality Control.
   (b) Total Quality Management
6. What are the sources of grievance in industry ? Explain the procedure to handle grievances.
7. Discuss financial and non-financial incentives as motivation. Give details of promotion policy and job satisfaction for motivation.
8. What is the role of advertising ? Discuss requirements of a good advertisement.

PART – IV

Project Planning and Control

Chapter

7

Project Planning

7.1 INTRODUCTION

In order to carry out work efficiently, it is essential that a scheme of operations be first decided by those directly responsible for the execution. With such planning, the work can be broken down into a series of operations and an orderly sequence or programme of execution evolved. Without a programme, execution can only be haphazard and disorderly.

The drawing up of a programme at the beginning of the work is not fixed. It is flexible and changes can be made during execution. The project management is based on the above concept. Project management also deals with proper identification, formulation and appraisal of the project. These three aspects form the basic foundation of successful project management.

7.2 PROJECT PLANNING

7.2.1 Planning

Project planning involves the following :

(*i*) Defining the output of the project in terms of product/services to be generated.

(*ii*) Forecasting and estimation of input resources for the project in terms of men, materials, money, machines, etc.

(*iii*) Breaking down the project into manageable work packages or activities and logical sequence between different activities.

(*iv*) Structuring an appropriate organization to implement the project.

(*v*) Planning the tentative project completion time by preparing a master programme schedule indicating control points and critical dates of major events.

(*vi*) Scheduling the activities to complete the project in minimum possible time by the use of network analysis.

(*vii*) Preparation of detailed cost estimates of all activities.

(*viii*) Estimation of required resources for all the activities.

(*ix*) Preparation of contingency plan considering all adverse circumstances.

7.2.2 Organizing

Organizing consists of the following sub-processes.

(*a*) **Financial Resources :** The financial resources are to be arranged to ensure their adequate and timely availability.

(*b*) **Project Organization :** A project manager must be appointed with responsibility for coordination, directing and controlling the implementation of the project. A suitable organizational structure is to be selected and manned with competent personnel with required skill and expertize.

(*c*) **Team Building :** Execution of project requires special and unique skills, capable personnel selected must be given training and counselling.

(*d*) **Vendors and Contractors :** The vendors for material supplies and contractors for execution of the project should be appointed following proper procedures.

(*e*) **Communication :** A suitable information and communication system should be designed and implemented to collect and disseminate all relevant information among members of the project team and other project stakeholders.

7.2.3 Executing

The project execution shall be done as per original plans. However, project is a dynamic process, flexibility is required in the execution of the project within overall objectives of the project.

7.2.4 Directing and Controlling

The project manager will direct his subordinates through the tools of instructions, orders, requests, guidance, supervision, coaching, advice, etc., and motivate them to achieve the project goals.

Controlling is checking the variances in time, resource and quality from original plans. After identifying the deviations, suitable remedial measures must be taken to correct the variances and put the project back on right track.

7.2.5 Closing

A project comes to an end when the execution is completed and project objectives are achieved. The formal process of closing the process consists of:

(*a*) Testing the project gives planned output.

(*b*) Settlement of final bills of the contractor.

(*c*) Information of closure is sent to all stakeholders, employees, executives and others involved in the project.

(*d*) Relocating men and material resources to other projects or departments.

7.3 THE FINANCIAL FUNCTIONS

An entrepreneur has to perform the following functions during planning and control of a project.

1. Estimation of Capital Requirement

An entrepreneur has to estimate the funds requirement for short-term and long-term investment in the project. The funds are required for the following functions.

(*i*) Purchase of fixed assets including land, building machines, fixtures, tooling, etc.

(*ii*) Meeting the working capital equipments including purchase of raw materials, salaries and wages of the employees, day-to-day expenses.

(*iii*) Modernization and expansion of the business.

2. Determination of Capital Structure

On the basis of above estimates, the entrepreneur has to devise the capital structure, i.e., proper mix of equity and debt and short-term debt-equity ratio and long-term debt-equity ratio. The capital structure selected should ensure minimum cost of capital and maximum wealth of share holders and entrepreneur.

3. Sources of Funds

The entrepreneur has to decide the sources for raising of funds. The various sources can be equity, preference shares, debentures of share holders, public desposits, term loans from financial and other lending institutions and loans from commercial banks for working capital.

4. Procurement of Funds

The procurement of funds depends upon cost of raising funds, market conditions, choice of investors, government policy, etc. The entrepreneur may have to issue the prospectus, negotiate with creditors and lending institutions.

5. Utilization of Funds

The funds available have to be properly invested in various assets in order to maximize the return on investment.

The principles of investment are :

(*i*) Safety

(*ii*) Profitability, and

(*iii*) Liquidity.

6. Disposal of Profits and Surplus Funds

The entrepreneur has to decide about the disposal of profits and surplus funds generated into two parts.

(*i*) Payment of dividend to the share holders.

(*ii*) Retention for business operations.

The disposal of the above funds depends upon the following factors :

(*i*) Trend of earning of the enterprize

(*ii*) Trend of market share of the enterprize

(*iii*) Funds requirement for future programmes

7. Cash Management

Sufficient funds must be available for purchase of materials and other current assets, payment of wages and salaries to employees and meeting day-to-day expenses for business operation. There should neither be shortage nor surplus of cash with the organization. It involves proper forecasting of cash inflows and outflows. Working capital management is most important function of the entrepreneur.

8. Financial Control

The entrepreneur is responsible for financial planning as well as financial control. Evaluation of financial performance is an important function of financial control.

The financial performance is measured by the following techniques :

(*i*) Return on Investment (ROI)

(*ii*) Budgetary control or control of capital expenditure

(*iii*) Cost control

(*iv*) Internal audit

(*v*) Break-even analysis

(*vi*) Ratio analysis

7.4 CAPITAL EXPENDITURES OF AN ENTREPRIZE

Following types of resources have to be arranged for launching a project. These resources constitute the project capital expenditures.

1. Physical resources.
2. Technological resources.
3. Human resources.
4. Financial resources.

7.4.1 Physical Resources

The physical resources required for an enterprize are as follows :

(*i*) Land, land development, boundaring, drainage, internal and approach roads, etc.

(*ii*) Factory buildings, auxiliary buildings and structures, substation, effluent treatment plant, water storage and treatment plant, material storage, cranes.

(*iii*) Infrastructures, water connection, power connection, communication, transportation and other facilities.

(*iv*) Furniture and fittings, lighting and ventilation, air-conditioners, heaters, computers, etc.

(*v*) Plant and machinery, spare parts, foundations, erection and commissioning, power connection, earthing, arrangements for maintenance.

(*vi*) Jigs and fixtures, tooling.

7.4.2 Technological Resources

Technical resources have to be selected to ensure high productivity and quality. Arrangements have to be made for the arrangement of the following technological resources.

(*i*) Product design

(*ii*) Technical know-how

(*iii*) Production processes

(*iv*) Quality control system

(*v*) Technical training of personnel.

The sources of the above technologies may be available indigenously or may have to be imported through foreign collaboration.

7.4.3 Manpower or Human Resources

The success of an enterprize largely depends on the quality of manpower. The human resources include managers, supervisors, engineers, skilled and unskilled labour force, marketing and other field personnel.

Following steps have to be taken to develop a good inventory of manpower.

(*i*) Employment of competent personnel

(*ii*) Manpower training

(*iii*) Effective leadership

(*iv*) Financial and non-financial motivation

(*v*) Appropriate organizational climate

(*vi*) Development programmes and opportunities

(*vii*) Empowerment and participation in management.

7.4.4 Financial Resources

Sufficient financial resources have to be arranged for the following purposes :

1. Fixed capital for arrangement of fixed assets.

2. Working capital for arrangement of current assets.
3. Capital for expansion and modernization programmes.

The main sources of funds are as follows :

1. Own funds contributed by the entrepreneur, promoters, share holders, profit invested.
2. Borrowed funds including debentures, loan; deposits and credits from public, banks, financial and other lending institutions.
3. Venture funds, seed money, risk capital.
4. Lease finance.

7.5 COST OF CAPITAL

The entrepreneur has to arrange sufficient amount of capital to meet various expenses for initiation of the project and continuation of business enterprize. Capital is required for investment in fixed assets and procurement of physical resources and technological resources for the project. Capital is also required to procure raw materials, human resources and to meet day-to-day expenses.

In case of small enterprize, the entrepreneur alongwith other promoters can provide funds from their own savings. But in case of medium and large enterprizes, funds have to be raised form various sources like general public, commercial banks, financial and other lending institutions, etc. The entrepreneur has to take the following decisions.

1. Requirement of long-term capital
2. Requirement of short-term capital
3. Cost of raising funds
4. Optimum capital structure
5. Return on investment, and
6. Appropriate time for raising funds from various resources.

7.5.1 Components of Cost of Capital

Broadly there are two components of cost of capital.

(*i*) Administrative Expenses for Raising the Capital

Cost of raising the capital from various sources can be some time huge. This may include printing of prospectus, advertisement, preparation of legal documents, mortgage agreements, brokerage, underwriting, commission, etc.

(*ii*) Rent for Using Capital

The person or institute providing the capital has to be paid the rent for using the capital. The rent may be in the form of interest, dividend or some other forms. The creditors have to be compensated as per rate of return on investment as per his expectation.

7.5.2 Factors Affecting Cost of Capital

The following factors influence the cost of capital :

(*i*) Sources of capital

(*ii*) Cost of raising the capital
(*iii*) Amount of capital
(*iv*) Capital structure
(*v*) Duration for which capital is required
(*vi*) Risk involved
(*vii*) Project completion time
(*viii*) Idle time of capital
(*ix*) Security against capital
(*x*) Time of repayment
(*xi*) Terms and conditions of repayment
(*xii*) Reputation and reliability of entrepreneur and other promoters
(*xiii*) Availability of alternative sources.

7.6 COST OF CAPITAL AND PROJECT PLANNING AND CONTROL

Project management is creation of an enterprize under the technical, legal time and financial constraints. The advent of capital intensive technologies has increased the importance of finance. Entrepreneur has to provide adequate finance for the launching and functioning of enterprize.

7.6.1 Financing of a Project

The following decisions have to be taken by the entrepreneur for starting a business enterprize.

(*i*) Estimation of capital requirement
(*ii*) Sources of raising capital
(*iii*) Scope of raising equity capital and venture capital
(*iv*) Availability of private sources of finance
(*v*) Availability of term loans from financial institutions
(*vi*) Financial support from government
(*vii*) Mode of procurement of assets.

7.6.2 Types of Financing Options

The following are the options available to the entrepreneur for financing a project.

(*i*) Personal Financing

An entrepreneur may use his personal savings for financing a project. He may also mobilise from friends and family members.

(*ii*) Public Financing

Public financing is raised through shares and debentures for which the company has to go public. This type of financing should be avoided if finance requirement is small.

(*iii*) Venture Capital Financing

Venture capital is the risk capital which is required to launch a new project. This is mostly availed by technocrats and scientists who cannot provide much of own capital but hope for high rewards from the project. Many venture capital institutions and private companies have been established in India. The capital is lent on the basis of merit of the project and entrepreneur.

(*iv*) Debt Financing

Loans are raised from commercial banks and other sources as per guidelines issued by the Reserve Bank of India. The loans are granted as secured loans and unsecured loans. Long-term loans are spread from 3 years to 10 years or more to purchase fixed assets. Short-term loans are spread over from a few months to one year to purchase current assets.

(*v*) Government Support

Both central and state governments have set up institutions to support small-scale industries. Government also provides financial support in the form of subsidies and tax benefits to enterprizes located in rural and backward areas.

7.6.3 Procurement of Assets

The entrepreneur has certain options, for procurement of fixed and current assets by alternative means.

1. Infrastructure

The entrepreneur needs land and building for factory premises and shops for sales outlet. An entrepreneur has the following alternatives :

(*i*) Purchase of land from a private party and construction of factory building.

(*ii*) Purchase of industrial shed in an industrial estate developed by the state government.

(*iii*) Arrangement of premises on rent or lease basis.

The decision will depend upon availability of funds, size and type of enterprize, location of enterprize, tax benefits and scope of appreciation of real estate. A similar decision has to be taken for shops for sales outlets.

2. Plant and Machinery

The entrepreneur has the following choices :

(*i*) Outright purchase of plant and machinery

(*ii*) Procurement of equipment on hire and purchase basis

(*iii*) Procurement of equipment on lease basis

(*iv*) Procurement of equipment on contract basis.

The entrepreneur can purchase some basic equipment and go for leasing of certain assets which are very costly and unproductive such as office equipment, computers, air-conditioners and furniture. Equipment used only seasonably may be obtained on contract basis. Certain types of works like market survey, advertising, retail distribution may be outsourced on contract basis.

3. Raw Materials

It is essential to fix optimum level of inventory to avoid blockage of cash. Material can also be purchased on credit basis.

All the above decisions and planning require different amount of funds and therefore, cost of capital is the key to project planning and control.

7.7 ECONOMIC EVALUATION

The financial analysis of a project looks at the project from a limited viewpoint only, *i.e.*, the revenues entering the project's own accounts. The banks will finance a project which ensures a high-priority use of a region's resources. The economic appraisal looks at a project form the viewpoint of the whole economy and social benefits to the region, state or the country. The economic aspects of a project can be :

1. Increased output of goods and reduction in prices.
2. Enhanced services and generation of additional incomes.
3. Increased employment and labour wages.
4. Layer Government revenues.
5. High earnings for the owners of capital.
6. Higher standard of living of society.
7. Increased national income.
8. Improved income distribution.
9. Foreign-exchange savings and inflows.

The project objectives should stress the above project benefits clearly.

7.8 RISK ANALYSIS

All projects are prone to some kind of risk as these are appraised on certain assumptions which are unavoidable. These assumptions can be on :

1. Project cost estimates
2. Life of project
3. Estimates of demand, production, sales and prices.
4. Political and social developments.
5. Changes in technology, prices and productivity.

An allowance is provided to take care of the above assumptions which is based on risk factors. This allowance can have decisive impact on the profitability of a project.

7.8.1 Factors Leading to Uncertainty

The following factors add to the degree of uncertainty :

1. The process or product may become obsolete.
2. There can be decline in the product demand.
3. Changes in government policy on business.
4. Price fluctuations.
5. Foreign exchange restrictions.
6. Inflationary tendencies.

7.8.2 Types of Project Risks

Some important project risks are as under :

1. Project completion risk
2. Resource risk
3. Price risk
4. Technology risk
5. Political risk
6. Interest rate risk
7. Exchange rate risk

1. Project Completion Risk

Completing the project in time and within estimated cost is a major achievement. Project time over-run results in cost over-run. If additonal funds are not pumped in, the project may come to grinding halt. Projects over-run can be due to bad management or technological obsolescence during implementation of projects of long gestation period.

2. Resource Risk

Shortage of raw materials, power, fuel, skilled manpower will jeopardize all profitability calculations as there will be reduction in capacity utilization, increase in production cost and reduction in estimated return.

3. Price Risk

Price fluctuations of both inputs and outputs can have adverse effect on performance of the project. The government interventions in price fixation and capability of competitors to sell the product at a lower price will affect the project performance.

4. Technology Risk

There are two types of technology risks which can adversely affect the project.

(*i*) Use of nonproven technology at commercial level.

(*ii*) Obsolescence of technology due to its rapid growth.

5. Political Risk

Political risk cannot be predicted easily. The government frames rules and regulation for regulating the country's economy.

The government intervenes in many forms :

— Levying and regulating taxes
— Regulating monopolistic trade practices
— Imposing import duties
— Promoting exports
— Prohibiting export of certain products
— Issuing import licenses
— Controlling foreign exchange transaction
— Price controls
— Expropriation
— Nationalization

6. Interest Rate Risk

Fluctuations in interest rates on long-term borrowings can adversely affect the project. Project implemented on high-cost borrowing cannot compete with new projects funded at lower rates. Increase in floating interest rates result in lower profit margins.

7. Exchange Rate Risk

Company exposed to international economy are adversely affected by volatile exchange rates. Exchange rate risk hedging tools such as forward cover, leads, and lags, currency options, currency swaps, etc. should be used against exchange rate risks.

7.8.3 Techniques of Risk Analysis

The techniques available for handling risk may be classified into two groups.

1. Conservative Methods

(*i*) Shorter payback period
(*ii*) Risk-adjusted discount rate
(*iii*) Conservative forecasts

2. Modern Quantitative Techniques

(*i*) Sensitivity Analysis
(*ii*) Probability Analysis
(*iii*) Decision Tree Analysis

1. Conservative Methods

(*i*) Shorter Payback Period : Projects with shorter payback period are preferred to project with larger payback period. It should be combined with "cut-off period" which denotes risk tolerance level in the organization.

(*ii*) Risk Adjusted Discount Rate : The minimum required rate of return is raised by adding "risk premium". The risk-adjusted discount rate is used for discounting under NPV or IRR. If risk-free discount rate is 10% and risk premium of 3% is added, then risk-adjusted discount rate of 13% will be used for NPV and IRR calculation.

(a) Advantages :

(1) Easy to understand and simple to operate.

(2) It adds an attitude towards uncertainty.

(3) It has appeal for decision-makers.

(b) Disadvantages :

(1) It is not easy to work out risk-adjusted discount rate.

(2) It is unscientific to use uniform risk discount factor for all projects.

(3) There is no premium for taking risks.

***(iii)* Conservative Forecasts :** A certainty equivalent coefficient is calculated by decision-makers subjectively or objectively and cash flows are reduced by the above factor.

$$\text{Certainty Equivalent Coefficient} = \frac{\text{Desirable Cash Flow}}{\text{Estimated Cash Flow}}$$

2. Modern Quantitative Techniques

***(i)* Sensitivity Analysis :** It is a simulation technique in which variables which may contain significant error in estimation are selected. The effect of errors of different sizes are calculated on the present value of the project. Some of the key variables are :

1. Costs
2. Prices
3. Project Life
4. Market Share

***(ii)* Probability Analysis :** A range of estimates alongwith associated probabilities are calculated and probability distribution curve is plotted. Two types of probability are used for decision-making :

(*a*) Objective probability is based on a very large number of observations under independent identical situations.

(*b*) Subjective probability based on belief of a person. For example : Capital expenditure decision which are not repetitve.

***(iii)* Decision Tree Analysis :** Decision tree is a graphical display of relationship between present decision and future events and future decisions and their consequences.

The major steps in a decision tree process are :

(1) The investment decision is defined clearly.

(2) The decision alternatives are identified.

(3) The decision tree graph indicates decision points, chance events and other data.
(4) It presents the relevant data such a projected cash flow probability distribution, expected present value, etc. on the decision tree branches.
(5) Choosing the best alternative by analyzing the results employed.

7.9 PUBLIC ENTERPRISES

Public enterprise is a form of business organisation owned and managed by centre, state government or any other public authority. The whole or most of the investment is made by the government. Public enterprize is autonomous or semi-autonomous corporation or a company established, owned and controlled by the state and engaged in industrial and commercial activities.

During 20th century various governments have taken active part in the industrial, agricultural, financial and commercial activities. In India, massive investments have been made in a number of sectors such as oil and gas, power, metals, banking, insurance, transportation, mining, railways and postal services and public enterprises occupy commanding heights in national economy. A large number of industries have been reserved exclusively for public sector. Such a list is enclosed in Annexure XIII. However, New Economic Policy (1991) has reduced the areas exclusively reserved for public sector and diluted the government stake. It has accorded full management and commercial autonomy to successful profit making public enterprizes and allowed to raise money from capital market and sell off or close down loss making units.

7.9.1 Characteristics

The chief characteristics of public enterprises are :

1. Autonomous or Semi-autonomous Organisation

It is an autonomous or semi-autonomous organisation. Some enterprises work under the direct control of the government and some organisation are established under statutes and Companies Act.

2. State Control

These organisations are financed, owned and managed by the central or state government.

3. Rendering Service

The primary objective of public enterprize is to serve the public at large by supplying the essential goods and services at a reasonal price and creating employment opportunities.

4. Useful to Various Sectors

These enterprizes serve all sectors of people without discrimation.

5. Monopoly Enterprize

These enterprizes enjoy monopoly in certain sectors such as Railways, Post and Telegraph and Energy production. Private sector is not allowed in certain sectors.

6. Use of Foreign Money

The foreign assistance received from industrially advanced countries is spent through public enterprizes for development of industries.

7. Public Accountability

These enterprizes are liable to general public for their performance.

8. Agent for Government Plans

The economic policies and plans of the government are implemented through public enterprises.

9. Financial Independence

These enterprizes are founded by government but enjoy financial independence by arranging finance for day-to-day operations.

7.9.2 Capital Expenditure

The board of the public enterprize is empowered to sanction capital expenditure within certain limits. The Public Investment Board plays a pivotal role in the appraisal and sanction of capital project.

The Planning Commission of Government of India has issued a Mannual on Feasibility Studies and suggests that the project should be appraised from technical, commercial financial and economic analysis.

The following sequence of presentation has been suggested :

1. General information on alternatives.
2. Preliminary analysis.
3. Project description.
4. Market analysis.
5. Capital requirements and costs.
6. Operating requirements and costs.
7. Financial analysis
8. Social profitability analysis.

The following methods of economic analysis have been emphasised.

1. Internal Rate of Return on Investment.
2. Sensitivity analysis for risk characteristics for the following cases :
   (*i*) Increase in all capital cost by 30 percent.
   (*ii*) Decrease in all labour costs by 30 percent.
   (*iii*) Reduction in raw material input cost by 10 percent.

7.9.3 Policies

Memorandum of Understanding (MOU) has been introduced between the government and public enterprize. It is a negotiated performance agreement between them. It specifies the intentions, obligations and mutual responsibilities of the two parties. It atempts to move the management by public enterprize from management by controls and procedures to management by results and objectives.

The memorandum of undertaking contain three sections :

1. **Section I** contains preamble, mission, objective in terms of growth, profitability, customers focus, people orientation, technology, images, etc.
2. **Section II** specifies quantitative and qualitative criteria for performance evaluation and targets.
3. **Section III** defines the delegation of powers to public enterprize and obligations/assistance from the government.

7.9.4 Practices

The following practices are followed :

1. Performance Evaluation System

The performance of public enterprize is evaluated as under.

(*i*) The quantitative and qualitative criteria for performance measurement are negotiated and spelt out.

(*ii*) Weights are assigned to criteria.

(*iii*) Performance targets are spelt out on a 5-point scale for each criteria.

(*iv*) Actual achievement on each criteria is measured and converted into a raw score.

(*v*) The raw score on each criteria is multiplied by the weightage assigned to the criteria to get the weighted achievement.

(*vi*) The weighted achievement on all criteria are added to obtain the composite score.

(*vii*) The overall performance is evaluated as follows :

Composite score	Performance evaluation
1.00 – 1.50	Excellent
1.51 – 2.50	Very good
2.51 – 3.50	Good
3.51 – 4.50	Average
4.51 – 5.00	Poor

2. Disinvestment

Serious concerns have been raised about the effectiveness and efficiency of some public enterprizes based on the above performance evaluation.

Disinvestment or privatisation has been carried out by transfer of ownership, partial or total, from government to individuals and non-government institutions.

The privatisation programme has been motivated by the following objectives.

(*i*) **Improvement in Efficiency** due to autonomy, accountability and employee pride in private sector.

(*ii*) **Generation of Resources** is very convenient in private sector.

(*iii*) **Promotion of Popular Capitalism** by broadening of base of equity of share holders by private sector.

7.9.5 Indian Experiment

Two broad approaches have been adopted in India for privatisation.

1. Strategic Sale

The controlling shake is sold to a single buyer. He brings in new technology, expertize and replaces the government and has the incentive to improve the governance and management of the enterprie.

2. Open Market Sale

The shares of the enterprize are sold to public at large in stages and disperses share ownership. This strategy has been found more appropriate for India as the government can progressively bring down its shake.

7.10 SWOT ANALYSIS

A company has to frame strategies to maintain or improve its competitive position in a dynamic environment to earn profits: In order to adopt to changing environment, an organization has to study the strength and weaknesses of its structure. SWOT analysis helps in project appraisal for expansion/diversification activities. A subjective estimate of strengths, weaknesses, opportunities and threats to the organization is carried out. Some-examples are given below:

1. **Strengths:** Customer goodwill, brand loyalty, financial resources, dynamic managers, entheuastic sales force, latest technology, etc.
2. **Weaknesses:** Obsolete technology, poor financial position, lack of trained human resources, old product design, inadequate research and development activities, etc.
3. **Opportunities:** Expanding market, government orders, export possibilities.
4. **Threats:** Increased competition, changing technology, economic and political uncertainity, internal labour disturbances.

Question Bank

1. Explain the various financial functions and their importance to small-scale industries.
2. Distinguish between "Ownership capital" and "Borrowed capital" ? What considerations are to be taken care of while determining debt equity ratio ?
3. Explain the different sources of financing project, along with their relative merits and demerits.
4. Write notes on the following :
   (a) Costs of capital approach to project planning and control.
   (b) Risk analysis for economic evaluation of a small project.
5. Explain the different resources to be arranged for launching a new enterprize.
6. What are the components of cost of capital ? What are the various factors influencing cost of capital ?
7. What is public enterprize ? What are its characteristics ?
8. Explain the practice of performance evaluation used in public enterprizes. What is Indian experiment of disinvestment and privatisation ?

❑❑❑

Chapter

8 Project Control

8.1 INTRODUCTION

Projects are becoming role difficult and innovative. Business is becoming more complex and there is a demand not for a product or service but a complete package to meet the entire need of the customer. The changes in the competitive environment have affected the projects. This has become a major source of competitive advantage. The speed of changes in technology and methods are increasing. Customer expectations of product and services are increasing all the time. The service sector is growing fast which is providing employment and economic growth. The cost of project, completion time and product quality and technical performance have to be controlled.

The project controlling may consist of the following :

1. Profit planning and programming
2. Planning of cash flow.
3. Operational expenditure.
4. Control of financial flows.
5. Design of control and communication systems.

8.2 PROFIT PLANNING AND PROGRAMMING

Gross profit is the difference between net sales and the cost of goods sold. It is the first and the broadest measure of profit. Operating profit represents profit from operations after considering the cost of goods sold and operating expenses. It reflects the profit generated by the normal and recurring business activities of the enterprize and does not take into account non-operating gains (or losses), interest expenses and taxes.

The net profit or net income or the bottom line is the profit after tax. It is a measure of the change in the owner's equity arising from the revenue and expenses of the accounting period.

Profit after tax = Net cash flow + Non-cash revenues – Non-cash expenses.

or, Proft after tax = Net-cash flow – Depreciation – Amortisation

8.2.1 Advantages of Profit Planning

Profit planning offers many advantages to a business enterprize. The various benefits of profit planning are as follows :

1. Performance Evaluation

The profit plan provides a continuing standard against which sales performance and cost control can be quickly evaluated.

2. Awareness of Responsibilities

Personnel are readily aware of their responsibilities for meeting sales objectives, controlling cost, etc. from the profit plan.

3. Cost Consciousness

Cost excesses can be quickly identified and planned, expenditure can be compared with budgets even before they are incurred, cost consciousness is increased, reducing unnecessary costs and overspending.

4. Disciplined Approach to Problem Solving

The profit plan permits early detection of potential problems so that their nature and extent are known. With this information, alternative corrective actions can be more easily and accurately evaluated.

5. Thinking about the Future

Most of the time, an entrepreneur neglects to plan ahead. As a result, opportunities are overlooked and crisis occur that could have been avoided. Development of profit plan requires thinking about the future so that many problems can be avoided before these arise.

6. Financial Planning

The profit plan serves as a basis for financial planning. With the information developed from profit plan, an entrepreneur can anticipate the need for increased investment in receivables, inventory, or facilities and need for additional capital.

8.8.2 Profit Programming

1. Project Profitability

(*i*) Manufacturing expenses = Cost of raw materials + Utilities + Manpower.

(*ii*) Selling and distribution expenses include commission payable to salesman + expenses towards advertisement (lumpsum) + publicity expenses.

(*iii*) Administrative expenses include expenses on postage, stationery, telephone and telegraph charges.

(*iv*) The interest has two parts: Interest on Term Loan and interest on working capital.

The interest on term loan can be 12.5% to 14.5% depending upon quantum of loan, lending agency and scheme of loan. The interest on working capital loan can be calculated @ 13.5% to 16.5%.

(*v*) A flat rate of 15% of value of machinery and 5% of value of building can be taken towards depreciation.

(*vi*) Miscellaneous expenses can be estimated on lumpsum basis.

(*vii*) Total cost will be sum of all above expenses.

(*viii*) Sales Revenue is already calculated.

(*ix*) Gross Profit = Total Sales Revenue – Total Cost.

(*x*) Tax should be calculated from current tax adviser.

(*xi*) Net profit = Gross Profit – Tax.

2. Profitability Calculations

The profitability is calculated as follows for three years.

S. No.	Description	1st Year	2nd Year	3rd Year
1.	Sales Revenue			
2.	Estimated Production Expenses			
3.	Production Profitability (1 – 2)			
4.	Other income			
5.	Profit before Tax (3 + 4)			
6.	Expected income tax to be paid			
7.	Net profit (5 – 6)			
8.	Net cash returns (7 + depreciations + investment allowance)			
9.	Percentage of net profit (*i*) On sales $\left\lvert \frac{(7)}{(1)\times 100} \right\rvert$ (*ii*) On investment			

8.3 PLANNING CASH FLOWS

Cash flows are the key element in the evaluation of project capital expenditures. The cash flow stream of a project has three basis components.

1. Initial Investment

It is the after-tax cash outlay or capital expenditure and net working capital when the project is set up.

2. Operating Cash Inflow

These are after-tax cash inflows resulting from the operations of the project during its economic life.

3. Terminal Cash Inflows

These are after-tax cash flow resulting from the liquidation of the project at the end of its economic life.

8.3.1 Project Life

The project life for estimation of cash flows is taken the minimum of the following.

(*i*) Physical Life of the Plant

The number of years the plant would perform the function for which it has been acquired. This concept is useful for determination of depreciation charge but not for making investment decisions.

(*ii*) Technological Life of the Plant

The period of time for which the present plant would not be rendered obsolete by a new plant. It is very difficult to estimate this life but an estimate has to be made.

(*iii*) Product Market Life of the Plant

The period for which the product of the plant enjoys a reasonably satisfactory market.

(*iv*) Investment Planning Life by the Firm

The period for which a firm wishes to use for investment analysis. It will depend upon the complexity and size of the plant. For small investments this period may be 5 years. For medium investment, it may be 10 years and for large investments, it may be 15 years.

8.3.2 Basic Principles

The following principles may be followed for estimation of project cash flows.

1. Separation principle.
2. Incremental principle.
3. Post-tax principle.
4. Consistency principle.

8.3.2.1 Separation Principle

There are two sides of a project, i.e., investment side (asset side) and the financing side. The cash flows associated with these sides should be separated. The cash flow or investment side is taken as rate of return on capital expenditure.

8.3.2.2 Incremental Principle

Find out the cash flows of the firm with the project and without the project. The difference between the two sides will give the incremental cash flows attributed to the project.

8.3.3 Post-tax Principle

Cash flows should be measured after taxes. The cash flow will be impacted by tax rates, losses and depreciation. The tax burden is progressive and proportional to total income of business, marginal tax rate should be used. Initially the firm may make loss which may produce tax savings. The depreciation charges produce tax benefits.

The Income Tax Act prescribes the depreciation rate as follows :

Land	: 0%
Building	: 5%
Plant and Machinery	: 25%
Computers	: 60%
Vehicles on hire	: 40%
Pollution control equipped	: 100%

8.3.3.1 Consistency Principle

Cash flows and the discount rate applied to the cash flows must be consistent with respect to the investor group and inflation. For estimation of project capital expenditure to the cash flow to all investors is applied, i.e, share holders and lenders. A nominal discount rate is used by incorporating the expected inflation.

Note : The cash flow forecasts in project evaluation are of critical importance, adequate care should be taken to avoid over-statement or under-statement of profitability of the project.

8.3.4 Cash Flow Calculations

S. No.	Description	1st Year	2nd Year	3rd Year
(*a*)	**Financial Inflow**			
1.	Net profit (income tax and interest added but depreciation substrated)			
2.	Own investment			
3.	Depreciation			
4.	Investment allowance			
5.	Long-term loan			
6.	Short-term loan			
7.	Other income			

Contd...

(*b*)	**Financial Outgo**			
1.	Preliminary and pre-operative expenses			
2.	Fixed investment			
3.	Increase in current assets			
4. 5.	Reduction in long-term loan Reduction in short-term loan			
6.	Interest			
7.	Income tax			
8.	Total			
(*c*)	**Repayment of Loans**			
1.	Net profit (income tax debited)			
2.	Entrepreneurs drawings			
3.	Available profit (1 – 2)			
4.	Amount to be returned to financial institutions			
5.	Interest on funds availed from state corporation and banks			

8.4 CAPITAL EXPENDITURE AND OPERATIONS

The operating expenditure is an ongoing cost for running a business. It includes cost of workers, rent, utilities, day-to-day expenses, etc. It is the money the entrepreneur spends in order to turn inventory into throughput. The operating expenses also include depreciation of plant and machinery used in production process, raw materials and purchased components. The operating expenses are also called working capital.

8.4.1 Management of Working Capital

Working capital can be Gross Working Capital or Net Working capital. Gross Working Capital is the total of current assets. Net working capital is the excess of current assets over current liabilities. It is a part of working capital, which is financed by long-term sources of funds, and this determines the margin of safety for meeting the current liabilities.

1. Significance of Management of Working Capital

1. Cost of working capital is higher because of higher interest rate charged by the bank on working capital loans than on term loans.
2. The fixed assets are purchased occasionally and therefore, the financial decisions are called for only occasionally. But management of current assets calls for decision-making on continuous basis and often involves huge amount of funds.
3. In an inflationary situation where prices of raw material show a constantly rising trend, unless working capital is managed efficiently; production can be hampered for want of funds to purchase raw materials.

2. Aim of Management of Working Capital

The aim of efficient working capital management is that the enterprise has sufficient funds to meet its daily expenses on production, sales and other operations and at the same time, the firm does not block its capital in maintaining inventory in excess of what is required, and that it is prompt in collecting its receivable. Blocking of excess capital results in higher expenses by payment of interest on capital. Therefore, there is decrease in the profitability. A firm must estimate its working capital need accurately and arrange outside borrowings.

3. Operating Cycle

Operating cycle is a circular flow in which cash is used to buy raw materials. Raw materials are converted to semi-finished goods and then finished goods by incurring the expenses of production. Finished goods are sold on to the customers on credit and the cash is blocked in the form of bills receivables. On realization of the receivables, the funds complete one cycle and become cash again to be used for purchase of raw materials. This cycle continuously goes on and therefore at any one point of time, the working capital employed by our enterprise is the sum of the following items.

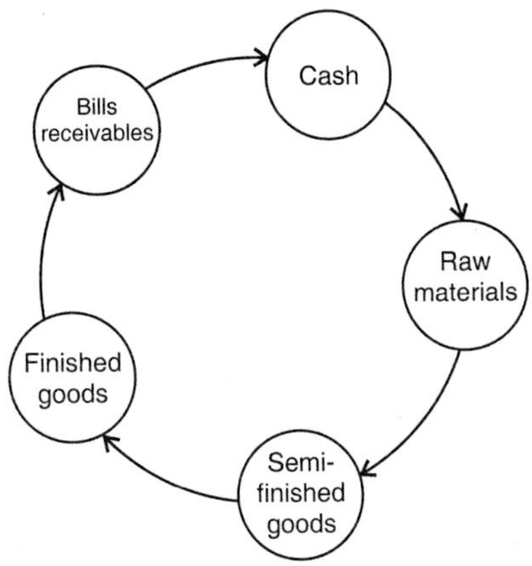

Fig. 8.1. Working capital operating cycle.

(*a*) Cash and Bank Balance.
(*b*) Value of Raw Material Stock.
(*c*) Value of Stock-in-process.
(*d*) Value of Finished Goods Stock.
(*e*) Value of Receivables.

The entrepreneur or his finance manager should design and control the quantum of above items scientifically.

4. Management of Cash

The size of cash balance to be kept depends on the following factors.

(*a*) *Volume of Transactions*

Some balance is to be maintained between the inflow of cash from sales realization and outflow cash in daily-operated expenses.

(*b*) *Contingency Needs*

Some amount of cash should be kept to meet the following contingencies.

(*i*) A major customer does not pay his bill,
(*ii*) A sales delivery lot is rejected due to improper quality,
(*iii*) A supplier suddenly reduces the credit period.

(*c*) *Opportunity Needs*

A supplier offers a large cash discount for purchase of raw materials.

Therefore, cash should be kept aside to make use of opportunities for financial gains.

(*d*) *Availability of Short-Term Credit*

If short-term credit is easily available from bank or money market, less cash balance can be kept.

(*e*) *Variation in Cash Flow*

Keeping harmony with the monthly cash flow projections different amounts of cash balance should be kept in different stages of a month. If there is any seasonability in the operations of the unit, it should be taken into account.

5. Management of Stock

(*a*) *Raw Material Inventory*

The volume of inventory will depend upon the following factors:

(*i*) Lead time of purchase, i.e., time lag between planning of order and actual receipt of materials.

(*ii*) Minimum order Quantity, i.e., the minimum quantity of a material is packed in or the suppliers executes an order.

(*iii*) Rate of consumption of raw material.

(*iv*) Ordering cost, i.e., cost of postage, stationery, telephone and visit to the supplier, which have to be incurred for placing an order. Higher the ordering cost, less will be the frequency of purchase and more will be the volume of one purchase. The maximum raw material stock will be high.

(*v*) Inventory carrying cost, i.e., interest, rent of godown, cost of insurance. If this cost is high, efforts should be made to reduce the volume of the stock.

(*b*) *Stock in Process*

The volume of stock-in-process mainly depends on:

(*i*) The period of completion of manufacturing process. The investment in stock-in-process will be higher where the manufacturing cycle time is long and vice-versa.

(*ii*) The stock-in-process is high when manufacturing process involves use of several machines and capacities and these machines are not balanced.

(*c*) *Finished Goods*

The volume of finished goods inventory depends upon the following factors:

(*i*) Units producing in anticipation of sale maintain higher stock of finished goods than units producing against orders.

(*ii*) Waiting lime for inspection specially third party inspection.

(*iii*) Waiting time for Railway Wagon/Road Transport facilities.

(*iv*) Batch size for delivery of goods. If batch size for dispatch is large, higher stock of finished goods is maintained.

(*v*) If optimum batch size of production is large, the finished goods stock will also be high.

6. Receivables

The volume of receivables depends upon:

(*i*) The payment terms of the customer and their liquidity position.

(*ii*) Credit policy of the selling firm. Selling in competition will require a liberal credit policy.

(*iii*) Cash discount. Some discount is allowed to the customer if goods are paid in cash.

(*iv*) High receivables involves blockage of funds and risk of debts going bad.

(*v*) Credit can be used as an investment for promoting sales.

8.5 CONTROL OF FINANCIAL FLOWS

Every small business struggles with the control of financial flows. Proper control of financial flows ensure success of a business. Entrepreneurs usually carry out financial flow projections, i.e., tabulation of expected revenues and expenses for the next few weeks or months. These projections are effective only if there are current and accurate. However, there is a powerful tool to monitor and control financial flows in manufacturing industries called Financial Flow Graph.

8.5.1 Construction of Financial Flow Graph

Financial Flow Graph is a running plot of cummulative orders, sales, cash in and cost out with time. A typical graph is shown in Fig. 8.1. The main advantage of this graph is that any changes in orders are reflected in cash-in, time-to-ship and time-to-collect later.

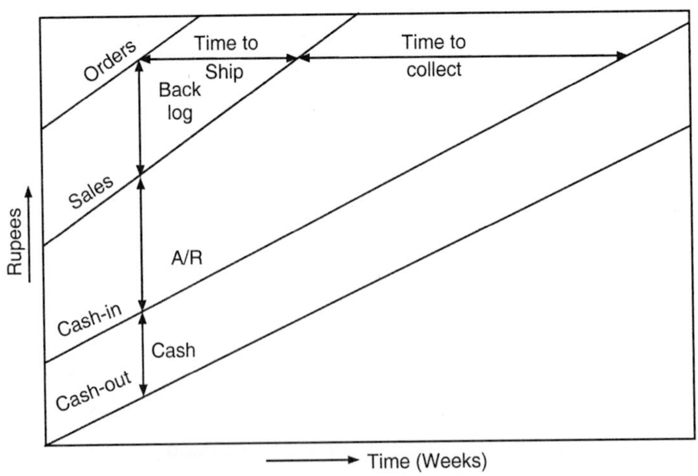

Fig. 8.2. Financial Flow Graph

For example, if the above times are 6 weeks each, there is a running 12 weeks advance knowledge of cash-in, i.e., plenty of time to take suitable action to avoid cash problems.

The Financial Flow Graph is constructed as follows :

1. Starting cash out = 0
2. Starting cash in = starting cash out plus current cash balance.
3. Starting sales = Starting cash in plus current accounts receivables.
4. Starting orders = Starting sales plus current back log.
5. Cummulative sales at n^{th} week = Cummulative sales at $(n - 1)$ week plus sales in n^{th} week.
6. Cummulative orders at n^{th} week = Cummulative sales at n^{th} week plus current backlog.
7. Cummulative cash in at n^{th} week = Cummulative sales at n^{th} week minus currents accounts receivable.
8. Cummulative cash out at n^{th} week = Cummulative cash in at n^{th} week minus current cash balance.

where

Sales at n^{th} week = Goods shipped and invoiced during n^{th} week.

Current backlog = Open orders to ship

Current accounts receivable (A/R) = Total of unpaid customer invoices.

Current cash balance = Total liquid cash.

8.5.2 Observations

Following observations are important for planning and control of financial flows.

1. Cash-in and cash-out lines should be diverging to achieve net profit.
2. Time-to-ship is a measure of resources in building and shipping the product, i.e., labour, materials, outside services, etc.
3. Time to collect is a measure of collection efforts.
4. If the order line starts curving upwards, time-to-ship will increase and to achieve accelerated sales, more resources have to be applied. Added resources have to be paid later.
5. In order to ensure continuity of accelerating orders, the cash-out line has to be changed by adding additional payments.
6. The projected cash balance should ensure that payments can be made for additional resources needed. If projected cash balance is not good, raise cash (equity or debt) to fund the growth.
7. Raise the additional cash at an acceptable cost.
8. The time-to-ship will automatically reduce the slope of order line and orders have to be cancelled. Avoid this reduction of orders to happen.
9. Some orders are more profitable than others. Therefore, cancel the last profitable orders so that more profitable orders can be serviced more adequately. This is called "skimming the cream" and is not an unethical business practice.

10. Therefore, the financial flow graph can be used to make following advantages.
    (*i*) It helps the entrepreneur to take proper decisions with best available data.
    (*ii*) Cash crises can be avoided well in time if timely decisions are taken.
    (*iii*) The graph gives plenty of advance warning if there are chances of cash crises.

8.6 CONTROLLING

Controlling is the process of checking existing operations against some desired results determined during planning process.

8.6.1 Components of a Control System

The essential components of a control system are :

(*i*) A predetermined goal, plan, policy, standard, norm, decision, rule, criterion, or yardstick.

(*ii*) A quantitative method of measurement of a current process to be controlled.

(*iii*) A method of comparison of current process with the criterion.

(*iv*) A method of correcting the current process to achieve the desired results.

8.6.2 Principles of Control Process

The following principles can be used to develop the control system.

(*i*) Optimum control can be achieved if critical points can be identified and close attention is directed to adjust these points.

(*ii*) Feedback system is required for the process of adjustment.

(*iii*) The control system must be flexible and must be responsive to changing conditions.

(*iv*) Control system must be designed to suit the organisation of the enterprize.

(*v*) There should be subsystems of self-control (by the unit or department) which can be integrated with and fitted to the overall control system of the enterprize.

(*vi*) The control system should be designed to maintain a direct contact between the controller and the controlled.

(*vii*) The control system must create a positive psychological impact on the human beings involved. A technically well-designed control system may fail due to unfavourable reaction of the people involved.

8.6.3 Applications

The control system can be applied to any personnal or managerial functions. Some examples are :

- Quality control
- Financial control
- Budgetary control
- Manpower control
- Project control
- Inventory control
- Marketing control
- Production control
- Cash control

8.7 COMMUNICATION

Communication is an essential ingredient in management-employee relations. It is the mechanism through which human relation have developed. Communication is the blood stream of an organization. It is the basic tool for motivation to increase the morale of the employees. Supervisor and leadership are impossible without communication. It is impossible to have human relations without communication and vice-versa.

The importance of communication in business can be judged from the following facts. Communication can help :

(i) Promote Managerial Efficiency

Manager deals with the personnel in giving orders and instructions, allocating job and duties, approving work, recognizing performance, explaining policies and objectives and seeking effective co-operation. An efficient manager has good communication ability.

(ii) Planning and Decision-Making

Facts, figures, information, issues and problems are communicated to manager continuously for planning and decision-making. Design of a good communication system is needed.

(iii) Strength Control and Operation

Conveying decisions, orders and instruction from top to bottom and bringing back reports, suggestions and response from bottom to top require efficient communication system.

(iv) Co-operation through Understanding

Effective communication creates the faith and results in satisfactory performance through greater understanding. Effective communication increases understanding.

(v) Leadership Action

Leadership is not possible without proper communication between the leader and his followers. Two-way personal communication is essential for maintaning man-to-man relationship in leadership.

(vi) Co-ordination

Communication helps in securing required coordination in enterprise operations by building communication networks throughout the organization.

(*vii*) **Job Satisfaction**

A clear-cut understanding between manager and employee provides job satisfaction. Mutual trust and confidence between manager and employees can be increased by communication.

Methods of Communication

The important methods of communication are:

(*a*) Verbal, (*b*) Written, (*c*) Gestural

Media of Communication are:

(*i*) Employee handbooks
(*ii*) Company periodicals, magazines
(*iii*) Employee's papers
(*iv*) Financial report to employees
(*v*) Documents covering personnel policies
(*vi*) Displays and notices on bulletin boards, notice boards
(*vii*) Audio-visual aids
(*viii*) Museum and exhibition
(*ix*) Posters

Effective Communication

(*i*) Send direct and simple messages
(*ii*) Feedback system
(*iii*) Using may communication channels
(*iv*) Face-to-face communication
(*v*) Message to fit the receivers vocabulary, interests and values
(*vi*) Time the message carefully
(*vii*) Reinforce the words with action
(*viii*) Introduce a proper amount of redundancy in message
(*ix*) Create cordial and peaceful atmosphere in organization

Question Bank

1. What is project control ? How can profit plan be used as a useful tool for project control ?
2. What are the main advantages of profit planning ? How is profit programming carried out ?
3. What is cash flow planning ? Discuss the principles of cash planning. How are cash flow calculations made ?
4. What is operating cycle used in working capital ? Discuss the management techniques used to control various components of operating cycle.
5. What is Financial Flow Graph ? How can this technique be used for effective control of financial flow ?
6. Write brief notes on the following as applicable to small business.

   (*a*) Controlling (*b*) Communication.

PART – V

Legal Obligations and Entrepreneurial Support System

Chapter

9 Legal Obligations

9.1 INTRODUCTION

An interpreneur will have to comply with some legal obligations during planning, implementation and operation of the enerprize. These acts and rules to be followed are called business or industrial laws. There are certain procedural requirements for setting up a small scale industry. Setting up an industrial enterprize consists of the fulfillment of a programme within the main constraints of technical, financial, time-schedule and legal obligations. Every industrial enterprize is expected to ensure compliance with various enactments, regulations, legislations etc. of central and state governments and those of local bodies before, in the cause of its establishment and after it is established.

During construction of the plant, some resources are to be arranged for which requirements of a number of institutions have to be fulfilled. Cement, steel, power, fuels, imports of equipment and raw materials and other controlled indigeneous raw materials, utilities and construction materials can be procured only by completing the requirements of a number of agencies.

Finance is the most difficult raw material to be arranged for which the requirements of the banks and financial institutions and other lending institutions have to be met. Information on various enactments may be available in different publications. The entrepreneur should know about the procedures to be followed.

9.2 LAWS FOR ENTREPRENEUR

The laws and statutory measures are intended to:

1. Promote regulated growth of business and industry.
2. Protect the interests of the consumer.
3. Prevent congestion or pollution in the cities.
4. Regulate employment of labour and avoid its exploitation.
5. Ensure proper working condition.
6. Promote harmovious relations between labour and management.
7. Provide measures of social security.
8. Regulate exports.

9. Ensure realization of foreign exchange and export of acceptable quality.
10. Protect the interests of prospective enterpreneur, and
11. Raise reveneues for the state.

All enterpreneurs fall within the ambit of one or many of these enactments. These enactments require the industrial enterprize or business organization to seek an approval, permit, licence or registration from the concerned authorities for forming a partnership or a company, setting up an industrial undertaking, aquisition of land, construction of factory premises, disposal of effluents or emission of smoke etc., and thereafter, to submit returns or reports to the competent authorities and maintain records in a prescribed manner, follow the procedures set out in the enactments while conducting the affairs of the undertaking. These enactments invariably prescribe penalties where the provisions are violated.

The important enactments may be grouped as follows:

1. Business organization
2. Industrial acts
3. Tax laws
4. Labour laws.

9.3 FORMATION OF BUSINESS ORGANIZATION

A business can be conducted by an individual as a propritor or in association as partnership, cooperative or joint stock company. The registration of partnership with the Registrar is optional, but its registration under Income Tax Act is desirable for lower rates of taxation for registered firms.

The private limited companies and public limited companies can start their business only after their incorporation. The conversion of a private limited company to a public limited company requires modification of memorandum and articles of association, broadbasing of Board of Directors, and listing of shares with Stock Exchange. The remuneration to the directors is controlled by Central Government. The amalgamation of companies requires concent of shareholders and creditors of both the companies being amalgomated and Income Tax Authorities. In case of dispute, the matter will be decided by High Court.

In addition to Companies Act, 1956, the company is controlled by various other acts like Capital Issue (Control) Act, 1947, the Monopolies and Restrictive Trade Practice Act, 1969, Foreign Exchange Regulation Act, etc.

9.3.1 Selection of Business Organization

The following factors should be considered for selection of a proper business firm.

1. Nature of Business

Proprietary concerns are best suited where personal attention and skill of entrepreneur is required. For pooling of funds and skills, partnership firms are suitable. Company form of organization is required where large-scale production is involved.

2. Area of Operations

A proprietor can handle local markets. However, a company only can handle national and international markets.

3. Degree of Control

Proprietorship or partnership firms can handle business with direct control whereas a company will be suitable for business not requiring direct control.

4. Capital Requirements

For business requiring small capital, a proprietorship or partnership may be appropriate whereas a company will be required if capital involved is large.

5. Risk and Liability

If entrepreneur is ready to take risk and liability of business, he can use proprietorship or partnership whereas if business involved is huge, a company should handle the business.

6. Time of Business

A permanent business should be run by a company or cooperative but business with limited duration should be handled by a proprietor or in partnership.

7. Government Regulations

A company or cooperative can handle business where many Government rules and regulations are involved. If Government involvement is less, select proprietorship or partnership firm.

9.3.2 Partnership Laws

In India, nearly 85% of small business is handled by sole propietors or Joint Hindu families, another 14% by partnership firms and nearly 1% small business is run by companies or cooperatives.

As per Indian Partnership Act, 1932, partnership is defined as the relation between persons who have agreed to share the profit of business carried out by all or any of them acting for all. The main characteristics, advantages and disadvantages of partnership firm are given under business ownership.

The salient features of Indian Partnership Act, 1932 are given as Annexure-VI.

9.4 BUSINESS OWNERSHIP

The important forms of business ownership organizations are as follows:

1. Sole proprietorship
2. Joint Hindu Family Firm
3. Partnership Firm
4. Joint Stock Company
5. Cooperative Society

The sole proprietorship, Joint Hindu Family Firm and partnership firms are non-corporate form of business organizations whereas joint stock company and cooperative society belong to corporate sector. The main distinctions between these two catagories are non-corporate organization can start without legal formalities whereas corporate organization can be launched and run after fulfilling certain legal formalities.

9.4.1 Sole Proprietorship

The sole owner arranges the capital, takes the risk of business and managers it. This type of ownership is easy to form and there is little legal formalities. This type of ownership is very suitable for service systems. The business can be easily taken care by one person. There is little risk of failure. This requires small capital to start the business and run it. Most of the service systems and small manufacturing systems are run by sole proprietors.

(*i*) Main Characteristics

The main characteristics of sole proprietorship are as follows:

1. Individual ownership.
2. One man effort.
3. Enterprise and owner entity is same.
4. All risks of the business are covered by one owner.
5. All profits after tax deduction belong to the proprietor.
6. There are minimum legal formalities to start and run the business.
7. Ownership and management have no separate entity.
8. There is unlimited liability. In case of loss or failure of the business, the entire debt or loan is recovered from the assets of the owner.

(*ii*) Advantages of Sole Ownership

Following are the advantages of sole proprietorship:

1. Simplicity of the organization.
2. Easy to establish, run and close the business.
3. Flexibility in management and easy to change the products or services, policies and control.
4. Easy decision-making.
5. High motivation to work for improvement and growth. All profits of business belong to the owner.
6. There is full control of business activities.
7. There is no compulsion to publish the accounts and the trade secrets can be easily maintained.
8. The operations are simple and less complicated.
9. There is wide potential for self-employment.
10. The business is passed on to the next generation in the family and therefore, there is indirect support for family employment.
11. This type of business is suitable for service systems and cottage and small manufacturing systems. The governments provide liberal

support in the form of easy loans from the banks, concessional water and electricity rates.

(*iii*) Disadvantages of Sole Ownership

The main limitations of this type of ownership are as follows:

1. There is limitation of resources depending upon the financial capacity of the owner.
2. In case of failure or loss in business, there is unlimited liablity of owner and the same is not covered by anybody else.
3. There is limitation of management skill as one man cannot have excellence in all spheres of production, marketing finance, liason, etc.
4. There can be discontinuity or closure of business in case of ill health or death of the owner.
5. There is always absence of specialized knowledge in one or more spheres of business.
6. The owner can be extremely stressed in complex operations or unseen situations.
7. There is chances of exploitation of owner by others in case of limited capital availability or high work pressure.
8. Due to limited hard and software resources, sole ownership is not suitable for large and complex operations.
9. There are limited checks and controls.
10. The business is less stable due to unavailability of another person to share the responsibility.
11. There is limitied scope of economics of economics of scale as the volume of production is less and cost of product or services can be high.

9.4.2 Partnership Firm

As per Indian Partnership Act, 1932, partnership is defined as the relation between persons, who have agreed to share the profit of business carried out by all or any of them acting for all. Partners, while joining the business, agree to share the capital resources and expenditure. In lieu, they share profit or loss as per the agreed proportion mentioned in the partnership agreement.

The number of owners are more than one but less than twenty. There is availability of more capital and assets and diversified expertise of the owners.

Partnership firms are suitable to carry out moderate size business in service systems as well as manufacturing systems. The typical examples are dealership, construction company, transport agency, automobile workshop, petrol pump ownership, small engineering firms, law firms, retail trade organization medical clinics, etc.

(*i*) Main Characteristics

The main characteristics of partnership firms are as follows:

1. The number of partners are more than one but less than 20. However, for banking sector, upper limit of partners is 10.

2. Partnership is governed by a partnership deed as it is a contractual formation of partners for carrying out a business. The agreement may be written, oral or implied.
3. The liability of partners is proportional to their share in the deed. In case of liquidation of firm, there is joint liability of all partners.
4. The deed becomes null and void in case of death or indisposition of any one partner. If required, a new deed is required to be enforced.
5. Deed is for profit in the business and not for charity.
6. The share of one partner can be transferred only after all other partners agree.
7. Each partner may act as representative or agent of the partnership firm.
8. There should be full trust and honesty among partners for the success of partnership. There should be full sharing of information about transactions in and fealing the clients.
9. It is not necessary to register the firm with the register. However, the registration of the firm helps to minimize the legal complications in case of disputes.
10. A firm can be dissolved any time on death of a partner or when all partners agree. Therefore, the duration of partnership firm depends upon the will of the partners.

(*ii*) Types of Partners

There are various types of partners based on extent of participation, profit and liability sharing.

1. **Participation:** There are active or working partners fully involved in the conduct of business. Sleeping or dormant partners invest in the business, collects his share of profit but does not take part in daily activities of management.
2. **Profit Sharing:** There are nominal partners who profit by lending their goodwill to the firm but do not invest in the capital nor take part in the daily management activities. There are partners-in-profit who do not take part in the daily activities nor share the loss. They have share in profit only. These partners are normally minors.
3. **Liabilities:** General partners cover unlimited liability irrespective of amount of contributed capital and take part in daily activities of the business. Limited partners cover limited liabilities to the extent of contributed shares.
4. **Conduct:** Estopped partner behaves as a partner of the firm in public, does not share profit but covers partial liability of the firm. Partner-by-holding out represents another person as partner of the firm.

(*iii*) Advantages of Partnership Firm

1. Partnership firm can be easily formed by an agreement or partnership deed.
2. There are a number of capital sharing partners who can pool large financial resources.

3. The various partners can share their managerial skills and capabilities.
4. The partners can take collective business decisions.
5. The interactive members are less and therefore, there is flexibility in change-over.
6. There is secrecy due to involvement of new members.
7. The partners have major share in the capital and direct risk in business dealing. Therefore, they take active interest in business.
8. All partners are very watchful and execute checks and controls.
9. In case of dissatisfaction, any partner can ask for dissolution of deed. The interest of partners is fully protected.

(*iv*) Disadvantages of Partnership Firm

Following are some limitations of partnership firm:

1. There is unlimited liability in case of dissolution of the firm.
2. As compared to a joint stock company, the capital formation is limited.
3. In case of death, bankruptcy or demand of a partner, there is risk and uncertainty of a continuation of the firms.
4. The firm can be put to loss by misdeeds of a partner and resulting loss has to be shared by all members.
5. Differences of opinion of partners can cause disharmony in operations of business.
6. The accounts and progress of the firm are not made public which can cause lack of confidence in public or institutions concerned.
7. The number of partners is limited to 20 causing difficulty in fund raising in future. Therefore, there can be difficulty in expansion and modernization of business.
8. A member can withdraw after the consent of all partners which may be difficult.

(*v*) Types of Partnership Firms

There are two types of partnership firms:

1. General Partnership
2. Limited Partnership

1. **General Partnership:** Each member has full executive powers, may act as on individual proprietor and can bind the partnership by an act. Additional members can be included with the consent of existing partners. The actions of one member can affect other members.
2. **Limited Partnership:** The firm has general partners as well as limited partners who have liabilities limited to the capital invested, can share profit but do not interfere with the management of the business.

9.4.3 Joint Hindu Family Business

The family business is governed by the Hindu Law and only the male members of the family can become the co-partner in the family. The property is inherited by the son, grandson and great grand son or by next three generations after the death of family head. The family business (or property) is handled by the *Karta* who is the head of the family. This system of inheritance is called *Mitakarshara* and is prevalent all over India except West Bengal where *Dayapbhaga* system is followed where both male and female members can become the copartners of the family business.

The salient features of the system are as follows:

1. Business membership is automatically granted afte the birth of a child.
2. The minors are full pledged members of the family business.
3. The lower limit on member of number is two with no upper limit.
4. There is no need for registration of family business.
5. The management and control of family business is handled by the head of the family (Karta).
6. Any member can ask for his/her share of account from the Karta.
7. The business is continuously run generation after generation.
8. The liability of Karta is unlimited whereas the liability of other members is limited to the share of their property.

9.4.4 Joint Stock Company

Joint stock company is an association of many persons who contribute money or money's worth to a common stock and employ it for common purpose. The common stock as contributed is denoted in money, and is a capital of the company. The persons who contribute it or to whom it belongs are members. The proportion of capital to which each member is entitled is his share. Shares are always transferable, although the right to transfer them is more or less restricted.

This form of business activity is most suited to large scale production system as there is no limitation of capital and management. The company can employ sufficient number of experts and skilled persons to run the business professionally.

(*i*) Main Characteristics

The company has the following main characteristics:

1. It is a voluntary association of persons.
2. It has separate legal existence.
3. There is a perpetual succession.
4. It is a registered body with common seal.
5. It is run by elected representatives known as directors.

(*ii*) Advantages of Company

The main advantages of company system are as follows:

1. There is limited liability of the members.
2. The shares are transferable.
3. The company can be managed efficiently and professionally by experts.
4. There is public confidence in the company and carries goodwill.
5. There is sense of social responsibility as social objectives are not ignored.

6. The ownership of the company is wide and distributed.
7. Large financial resources can be collected.
8. Company has good stability.
9. The management has the potential for growth and expansion of the company.
10. The company can opt for large scale business and take the advantage of economics of scale.
11. Government give various tax reliefs from time to time.

(*iii*) Disadvantages of a Company

There are certain limitation of a company:

1. There are legal formalities for the formation of the company which are time consuming.
2. The share holders lack motivation.
3. The board of directors can misutilize the capital of the share holders by corrupt practices.
4. There can be delay in decision making due to red tapism and bureaucracy.
5. There can be lack of sense of responsibility due to absence of personal intiative.
6. There is excessive government control and regulations inspite of recent openness and globalization.
7. The directors may manipulate book of accounts for personal gains at the expense of interest of share holders.
8. There can be heavy speculations in the share market and many share-holders may suffer badly.
9. The voice of share holders may fail to control the misutilization of resources as directors may mislead the share holders by controlling the business.
10. The board of directors take all the major decision on behalf of the share holders, whose capital is booked in the business.

(*iv*) Types of Companies

The company may be classified on the following basis:

1. Incorporation: (*a*) Chartered company, (*b*) Statutory company, (*c*) Registered company.

2. Liability of Members: Limited company, unlimited company.

3. Ownership: Public limited company, private limited company or Government company.

(*a*) **Chartered Company:** These companies are established by special sanction of the head of state of Royal Charter. The chartered company is granted exclusive powers and privileges. East Indian Company was eastablished by charter of Queen of England and Bank of England. These types of companies are not found these days as monarchy is almost over in the world.

(*b*) **Statutory Company:** These companies are formed by a special but separate act of parliament/state assemblies. All powers, objectives and rules are specified in a special act. Reserve Bank of India, Unit Trust of India, Industrial Development Bank of India belong to this category of companies.

(*c*) **Registered Company:** These companies are registered with the registrar of companies under companies Act and are governed by Company Act. TISCO, Reliance Industries Ltd., are registered companies.

The registered companies are classified as unlimited company and limited company based on the liability of the members.

(*a*) **Unlimited company** is not very common where the liablity of the members is unlimited and personal property of the members can be attached to meet the obligation of the creditors for repayment of loans.

(*b*) **Limited company** is registered with a specified capital being contrubuted by the members. The liability of a member is limited to the number of shares he holds. However, there are guarantor members who guarantee a limited amount over and above their share capital which is called upon only at the time of winding up the company.

4. Ownership of Registered Company: The companies are classified as follows based on ownership:

(*a*) Private limited company
(*b*) Public limited company
(*c*) Joint sector company

(*a*) **Private limited company** has the following main features:

- It is established by an Article of Association.
- It is registered with the registrar of the companies.
- The number of members can be between 2 and 50.
- There is restriction on the right of member to transfer the shares.
- Only members can subscribe capital to the shares and debentures.

(*b*) **Public limited company** has the following mains features:

- It is established by an Article of Association.
- The company is registered with the registrar of the companies.
- There is no upper limit on the number of members.
- Company can invite public to subscribe to shares and debentures.
- There is no restriction on the right of the member to transfer shares.

(*c*) **Joint sector company** is a registered company where minimum 51% of share capital is held by state and/or central government.

9.4.5 Cooperative Organization (or Society)

Cooperative organisation is an association of persons, usually of limited means who have voluntarily joined together to achieve a common economic end through the formation of a democratically controlled business organization, making equitable contribution to the capital required and accepting a fair sharre of risks and benefits of the undertaking. It is formed under the Indian Cooperation Societies Act, 1912.

(*i*) Main Characteristics

The main features of a cooperative society are as follows:

1. It is a voluntary organization.
2. It is formed by relatively economically weaker sections of society.
3. There is open membership.
4. The main objectives of the society is service motive, mutual help and for common interest of the members.
5. It works on the principle of one person–one vote and has democratic organization.
6. Registration of society is compulsory.
7. It has separate legas entity.
8. The profit, interest or surplus capital is disbursed among members in accordance with their share capital.

(*ii*) Advantages of Cooperative Society

The following are the merits of cooperative organization:

1. The cooperative societies can be **formed** easily with minimum 10 members and no upper limit of membership. The society has to registered with the Registrar of the Cooperative Societies.
2. **Any person** irrespective of caste, creed, gender or religion can become the member of the society.
3. It is **voluntary organization** without any compulsion of membership.
4. Cooperative society has **full autonomy** to invest capital or do any business which is legal and beneficial to every member.
5. Cooperative society has separate legal entity which ensures **stability and continuity** in the event of insolvency or death of any member.
6. The members have **limited liability** to their contribution and are not responsible for any debt of the society.
7. The election, functioning and management of the society is **democratic** with one member one vote principle.
8. The surplus and income are distributed among members and additonal shares may be allotted from the profit generated for additional **capital generation.**
9. Government gives low interest loan, special quota in land and housing and other **patronages** to the societies.
10. Cooperative societies are not involved in speculative business and there is **limited risk** to the members.
11. Cooperative societies are very suitable for weaker sections of the society as the main objective of the society is the **welfare** of the member.

(*iii*) Disadvantage of Cooperative Societies

Cooperative societies suffer from the following limitations:

1. A limited capital is generated through shares and subscriptions of limited members and business activity is llimited to **small size.**
2. The members elected to run cooperative societies may lack **management competency.**
3. **The commitment and interest** of the members in the normal functioning of the society may be lacking.
4. Cooperative society is a part-time activity and may suffer from **lack of coordination** among members.
5. There is lack of interest in running the society due to **low return** on investment. The members are generally interested in annual general meetings with lunch/dinner.
6. There is **no secrecy** of the business activities as every activity of the society is exposed to the members.
7. There are **chances of corruption** due to manipulation of accounts and audits due to limited interest by majority of members.
8. There is lack of regular activities, proper communication and clash of interest resulting in shift **among members.**
9. The shares of the society are **non-transferable** and it may be difficult for a member to quit the society.
10. The cooperative society is registered with the government which may use undeu and **unjustified controls.**

(*iv*) Types of Cooperative Societies

The cooperative societies can be classified as follows based on their purpose of formation:

1. Consumer's Cooperation Society.
2. Industrial Cooperative Society.
3. Cooperative Marketing Society.
4. Cooperative Credit Society.
5. Cooperative Housing Society.
6. Cooperative Financing Society.

1. **Consumer's Cooperative Society** is formed to eliminate middleman between consumers and producers, to ensure steady and regular supply of goods and services to the consumer and to share the profit in the form of dividend to the members.
2. **Industrial Cooperative Society** is formed to supply raw materials, tools and equipment to small-scale production organization and artisans to increase productivity, the goods are collectively produced and sold to face competition and the income is distributed among members proportional to amount of goods sold by the members.
3. **Cooperative Marketing Society** is formed by small producers and artisans to secure appropriate price for their goods sold, to improve bargaining power in the market, to place the small producers and

artisans in a competitive position, to eliminate the middleman in marketing of goods, to provide financial support to the members, to save the small producers and artisans from excessive financial and resource constraints in marketing of goods, by providing common transportation, warehousing, grading, packing, market research, market support and actual marketing of goods.

4. **Cooperative Credit Society** is formed to promote the habit of saving among its members, to provide credit or financial assistance as per need and to save the members from the exploitation by money lenders.
5. **Cooperative Housing Society** is formed to buy land from municipal and other authorities, construct houses or flats for the members and provide residential accommodation to the members on ownership basis on easy installments.
6. **Cooperative Financing Society** is formed for small farmers, to provide seeds, irrigation tools, fertilizers and other agricultural inputs to the members and assist in scientificfarming and mechanization to increase productivity.

9.4.6 Comparison of Different Forms of Business Ownership

The main characteristics of various forms of ownership of production systems are tabulated in Table 9.1.

9.5 INDUSTRIAL ACTS

The entreprneur has to design, construct and operate his enterprize to essure safe and healthy working conditions for his employees. There are a large number of industrial legislations depending upon the line of business. However, important industrial laws are as follows :

1. Factory Act, 1948.
2. The Indian Boiler Act, 1923.
3. Electricity Safety Act, 1910.
4. Smoke Nuisance Act.
5. The Water (Prevention and Control of Pollution) Act, 1974.
6. The Air (Prevention and Control of Pollution) Act, 1981.
7. Patent Rules.

9.5.1 Factory Act, 1948

The working conditions in a factory should ensure health and safety of the workers and other employees. The Factory Act, 1948 is a comprehensive document covering various aspects of design and operation of factories and regulates the working conditions. The Act extends to all part of India.

Factory Act Covers the following:

1. Licensing and Registration
2. Health
3. Safety
4. Welfare
5. Hours of work
6. Employment of young persons.
7. Leave.

The salient features of Factory Act are given as Annexure I.

TABLE 9.1. Various forms of ownership

S. No.	Characteristics	Sole proprietorship	Joint Hindu family	Partnership	Private limited company	Public limited company	Cooperative society
	1	2	3	4	5	6	7
1.	Ownership	Single	Family	More than 2 less than 20	More than 2 less than 20	More than 7 no upper limit	More than 10 no limit
2.	Formation	No legal formality easy	No major legal formality easy	No major legal formality moderately easy	Different legal formalities	Major legal formalities	Moderate legal formalities
3.	Separate legal	None	None	None	Yes	Yes	Yes
4.	Capital required	Very less	Limited	Limited	Large	Major	Net substential
5.	Management	By owner	By owner	Shared by owner	Owner or hired experts	Separate	Few elected members
6.	Security	Complete	Complete	Shared among partners	Shared among members	No	No
7.	Registration	No	No	Yes, fairly low	Yes, fairly high	Yes, highly regulated	Yes, moderate
8.	Expertise needed	Limited	Limited	Limited	Fairly high	Quite high	Moderate
9.	Liability of	Unlimited	Unlimited	Unlimited	Limited	Limited	Governned by By-laws

Contd...

	1	2	3	4	5	6	7
10.	Profit sharing	Compute by owner	Complete by owner	Shared among members	Proportion to held	Proportion to held	Based on business Volume by members
11.	Governing law	No	No	Indian Parliament Act, 1932	Companies to Act, 1956	Companies Act, 1956	Cooperative society Act, 1912
12.	Transfer of ownership	Any time	To son after death of father	Relatively difficult	Relatively difficult	Very easy transfer of shares	Restricted
13.	Tax structure	Very less	Very less	Very less	Heavy	Heavy	Exception
14.	Audit	No	No	No	Must	Must	Must
15.	Closure	Any time	Any time	Any time	As per Act	As per Act	As per Act
16.	Documen-tation	Not much writ-ten papers	Governed by will of owner	By consent of partners	Not much	Governed by memorandum of association systematic	Governed by memorandum of association very systematic
17.	Suitability	Life of owner	After death of owner passed to son	Depends upon will of all partners	Continuous	Continuous	Continuous
18.	Suitability	Small business as per expertize of owner	Small business depends on appertie of owner	Medium business depends on capi-tal, appertize and market forces	Medium to large business, depends upon market forces	Medium to large business depends upon govt priority	Medium business depends upon mutual benefits of members

The factory plans are submitted to Factory Inspector. The Chief Inspector of factories approves the factory plans as per Factory Act and issues licence of manufacture. There are pereodic inspections by the Factory Inspector to see that the Factory Act and Regulations are strictly followed.

9.5.2 The Indian Boiler Act, 1923

Every owner of the boiler is required to register his boiler with the Chief Inspector of Boilers and submit drawings of boiler. The inspector inspects the boiler and tests it hydraulically. He inspects the material and installation of equipment, piping mountings and accessions at site and give approval for boiler firing. The salient features of Indian Boiler Act are given in Annexure-II.

9.5.3 Electricity Safety Act, 1910

The plan of Substation installation of equipment and wiring of substation and factory are approved by the Electrical Inspector as per Electricity Safety Act, 1910 and State Electricity Rules: There is periodic inspection by the Electrical Inspector during production.

The Salient features of Electricity Safety Act, 1910 are given in Annexure-III.

9.5.4 The Water (Prevention and Control of Pollution) Act, 1974

Every industrial entriprise has to file all data pertaining to the assesment of water pollution sources, quantity of westewater generated and treatment plant as per prescribed proform to the Water pollution Board. The Board after detailed securitiny issues a "No Objection Certificate." This is required before acquisiction of land and construction of factory premises and allied buildings.

There is periodic inspection by the Board during plant operation to control that the instructions are fully implemented.

The salient features of the Water (Prevention and Control of Pollution) Act, 1974 are given in Annexure-IV.

9.5.5 The Air (Prevention and Control of Pollution) Act, 1981

Air pollution means any solid, liquid or gaseous substance (including noise) present in the atmosphere in such a quantity as may be injurious to human beings, other living creatures, plants or property or environment.

Every industrial enterprize has to file all data pertaining to the assessment of air pollution as perproforma given in the Act. Emission inventory of air pollutants from melting furnaces, boilers stock data and details of emission control equipment including calculations of height of the boiler chimney or stack has to be submitted to Air Pollution Control Board for approval and issue of "No Objection Certificate" before acquisition of land and construction of factory premises.

There is periodic inspection by the board during plant operation to control that the instructions are stricIty followed. The salient features of the Act are gives in Annexure–V.

9.6 PRINCIPLES OF TAXATION

Business enterprises and the businessman are subject to several taxes. Taxation is required to raise revenue to meet government expenditure in the following functional areas :

1. Administration
2. Defence against external aggression
3. Police for maintenance of law and order
4. Judicial courts for administration of justice
5. Education of population through schools and colleges
6. Health care of the public through hospitals.

The Public Finance consists of collection of revenue and its expenditure by the Government.

The general principles of taxation are:

1. Taxation should be equal, *i.e.*, proportional to their income and people should feel the burden of tax equally.
2. Taxation should be certain. The form, quantity and manner of payment of tax should be plain and clear to the contributors.
3. Taxation should be timely and there should be convenience of payment and collection of the tax.
4. Taxation should be economical to collect. Cost of collection of small incomes can be heavy.

Direct and Indirect Tax

Taxes may be Direct or Indirect.

1. Direct Tax

It is imposed on persons who are desired and intended to pay. For examples, Income Tax, Corporation Tax (Profits of companies), Capital gains tax, etc. The tax payer knows what to pay, why to pay and when to pay the direct tax. Direct taxes are collected at the source of income. The cost of collection is very small and chances of erosion are little.

2. Indirect tax

An indirect tax is passed on by one person on whom it is imposed to other person. These taxes are levied on goods and services when purchased and are called outlay taxes. The amount of tax depends upon the use of the taxed goods or services. Examples are : sales tax, excise duties, local tax, professional tax, service tax, etc.

Indirect taxes make no distinction between the rich and poor. The principle of direct taxes is a spread taxation over as wide a range of goods and services as possible.

9.7 INCOME TAX

Income tax is the tax levied on income as defined by Income-tax Act, 1961. It is an annual tax collected by the Central Government for each financial year on the total income of the previous year. The financial year of the Central Government is from 1st April to 31st March and, therefore, taxes have to be paid for each assessment year on the income of the previous year relevant to it. The Income-tax rates are decided by the Central Government every year while presenting the budget. The important points to be known are as under:

1. Accounting Year

A businessman has a freedom to choose his accounting year. He can keep his accounting year ending in any of the Christian calendar year or Vikram Samvat year. Once the choice is exercised, the change in the accounting year cannot be made without the permission of the Income-tax officer.

2. Status

The entrepreneur has to understand his legal and residential status. There are different rates of income tax for different status.

The legal status are:

(*i*) Individual
(*ii*) Hindu Undivided Family
(*iii*) Company
(*iv*) Partnership Firm
(*v*) Association of persons.

Residential status are:

(*i*) Residential (Resident and ordinarily resident)
(*ii*) Resident but not ordinarily resident
(*iii*) Non-Resident (NRI)

The entrepreneur must know the implication of his status on the rates of tax, exemptions, deduction vary with the legal status of the assesses.

3. Business Income and Deductions

Income-tax is levied on the income as defined by the Income-tax Act. There are various kinds of exemptions, deductions and incentives given by the government to reduce tax burden.

These deductions depend upon:

— Status of the tax payer (Individual, Company, Partnership)
— size of the industry (small-scale)
— nature of industry (Poultry farming)
— location of industry (Backward district)

These exemption keep on changing. The entrepreneur in consultation with tax expert, should plan his tax liability to be reduced to legally minimum possible.

9.7.1. Computation of Business Income

For computation of business income, the profit from the business or profession should be determined and adjustments should be as per provisional of the law.

(*A*) Deductions expressly allowed in computing business income

(*i*) Rent rates, taxes, repairs and insurance of buildings used for business purposes.

(*ii*) Current repairs and insurance of machinery, plant and furniture used for business purposes.
(*iii*) Depreciation of buildings, plant and machinery, furniture and fittings owned by the assessee.
(*iv*) Investment allowance in respect of plant and machinery.
(*v*) Expenditure on scientific research.
(*vi*) Expenditure on patent and copy rights.
(*vii*) Amortisation of preliminary expenses.
(*viii*) Insurance premium against risk of damage or destruction of stocks.
(*ix*) Bonus and commission to employees.
(*x*) Interest on borrowed capital.
(*xi*) Contribution to provident fund.
(*xii*) Bad debts written off.
(*xiii*) Any other expenditure (not being capital or personal expenditure) which is incurred wholly and exclusively for purposes of business or profession.

(B) Amounts expressly prohibited as deduction

(*i*) Any tax paid in proportion of profits, e.g., income-tax.
(*ii*) Wealth tax.
(*iii*) In case of firm, payment of interest, salary, bonus, commission on remuneration to partners.
(*iv*) Expenditure in the nature of charity or donation.
(*v*) Capital expenditure.
(*vi*) Personal expenses.
(*vii*) Amount transferred to a Reserve Account.

If the net result of computation is a loss, the source is allowed to be set-off against the income of any other head of income for the year. If it could not be so set off fully, this balance loss can be carried out forward to 8 subsequent years and set off against the business income of those years.

9.7.2. Deductions

Deductions are in the nature of relief or concession and given as an incentive to savings and investment. Such deductions are allowed from Gross Total Income to determine total taxable income.

Deduction are of two types:

(*a*) in respect of payments; and
(*b*) in respect of income.

Deduction on payments

(*i*) Life insurance premium, contribution of PF and investments in approved securities.
(*ii*) Donations of charitable institutions.
(*iii*) Medical expenses of handicapped parents.
(*iv*) House rent paid.
(*v*) Donation for scientific research or rural development.

Deduction on income

(*i*) Dividend and Bank interest
(*ii*) Lottery winnings
(*iii*) Long-term capital gain in case of non-company assess.
(*iv*) Part of profits of new industries undertaking or hotel establishment in backward years for a number of years.
(*v*) Part of profits of new small-scale industries in certain rural areas for a number of assessment years.
(*vi*) Part of profits on export turn over and Net Foreign exchange Realisation.
(*vii*) Part of profits relating to execution of foreign projects.
(*viii*) Part of profits of new industrial undertaking or a shop or a Hotel.
(*ix*) Certain income of co-operative societies.

9.7.3. Tax-payer's Obligation

The following obligations have to be met by the assessee under the tax laws otherwise he is exposed to serve penalties and prosecution.

1. Maintenance of Account Books

Account Books have to be maintained to enable correct computation of income. These books have to be preserved for a number of years.

Similar rules are applicable to a person engaged in legal, medical, engineering or accountancy profession.

2. Permanent Account Number (PAN)

Any person whose income exceeds maximum amount not chargeable to tax or whose business turnover exceeds a limit, must apply in Form 49-A allotment of PAN.

3. Income Tax Return

All persons whose income exceeds the excemption limit must file their income tax return in prescribed applicable form in time before a due date.

4. Payment of Tax

Taxes have to be paid within due dates. He should comply with notices issued by Income Tax Officer and notice may call for production of account books, bank accounts and other evidence to support the return filed.

9.8 EXCISE DUTY

Excise duty is a tax on the production or manufacture of excisable goods produced or manufactured within the country and is unrelated to the sale of the same. It is an indirect tax and is levied on the manufactures who passes on to the customer as part of price of goods sold. Ultimately its incidence is always on the consumers. The excise duty is calculated on the basis of Central Excise Act, 1944.

Types of Central Excise Duties

1. Basic excise duties

These are levied on a wide range of commodities classified under the schedule to the Central Excise Tariff Act, 1985. The basic law is under Central Excises and Salt Act, 1944. The basic duty can be specific, advolorem or a fixed tariff value.

2. Special duties of excise

These are levied under annual Finance Act expressed as %ge of basic excise duties (item 1).

3. Additional excise duties in lieu of sales tax

Textiles, sugar, tobacco, etc. are items on which additional duties are levied under Additional Duties Act, 1957.

4. Additional duties leviable on specific items under different enactments such as Medical and Toilet Preparation.

5. Cesses of excise duties leviable on certain specified commodities under various Acts. The products covered are Vegetable oil, jute, automobiles, etc.

Excise is a tax levied by the central government on the clearance of goods from a manufacturing factory. The rate of excise depends upon items manufactured and these vary as per the decision of the government. The penalties under the Excise act are very severe. A businessman, therefore must know the following:

(*i*) He must satisfy himself whether his product is free from excise or it is excisable item.

(*ii*) He must know the rate of excise leviable when the product exceeds the exemption limit.

(*iii*) The Government gives set off on the purchases made by the enterprise, if an excise duty has been paid by the supplier. This is very important because supplies from one source may give him set off, whereas for purchases from other source, such set off may not be available.

(*iv*) A businessman has to get a licence, maintain all clearance records, system of gate passes, etc. A businessman must follow these procedures.

An entrepreneur must consult an expertin in Excise Rules as these rules are very complex, change frequently.

9.9 SALES TAX

Sales tax is a state lovy made applicable whenever the goods are purchased from within the state.

When the goods are purchased from outside the state, Central Sales Tax (CST) is levied. The sales tax has to be charged by the seller from the purchaser and then handover the same to the government.

The businessman must acquaint himself with the following guidelines.

1. Basic Scheme

A firm with turn over exceeding a limit has to get sales tax registration with the state government. It is optimal for the firm to get CST registration. The rates of CST are considerably lower than those of state sales tax, registration of CST would give significant economic advantage. A registered business is called registered dealers (RD). The following facts must be known:

(*i*) Registration is compulsory or not.

(*ii*) What is the sales tax on the product.

(*iii*) Form 'C' is required for Central Sales Tax.

(*iv*) If he is selling outside the state, he will collect 'C' form from authorised dealer.

(*v*) A purchased product can be sold. A resold product may have zero or lower sales tax.

(*vi*) Different incentives given by government.

The sales tax has been replaced by uniform VAT (Value Added Tax) throughout the country.

2. Procedure

(*i*) **Registration of the Business under sales tax law.** Application on prescribed form has to be made to the Sales Tax Department in the state enclosing copies of partnership deed, proof of permanent place of business (Municipal Tax Bill). The fees required are nil. The department will issue Sales Tax Number.

(*ii*) **Filing of Returns**

The various steps required are:

(*a*) Registration as dealer.

(*b*) Maintenance of books of accounts.

(*c*) Filing of period returns showing all sales and purchases; taxable balance of amount of tax to be paid.

(*d*) The tax payable to be deposited in the treasury (through authorised bank) with the challan in a prescribed form.

The entrepreneur can take the help of an expert.

3. Effect of sales Tax on Sales Price, Cost and Competencies

(*i*) Know the exact rate of sales tax on raw material and finished goods.

(*ii*) Benefit or incentives declared by state government.

The product may not attract Central Excise but sales tax is always leviable. his will raise the sale price. Also the cost of raw materials get raised after paying sales tax on the purchases. The sales price would be higher. The sales price can be higher due to higher state sales tax as compound to the price of similar product in the neighbouring state. Sometimes an imported item may be cheaper. Due to higher rate of sales tax, you may lose competition.

9.10 GOODS AND SERVICE TAX (GST)

9.10.1 Definition and Scope

"Goods and services tax" means any tax on supply of goods, or services or both except taxes on the supply of the alcoholic liquor for human consumption.

1. Goods 366 (12) includes all materials, commodities, and articles [Pre Existing Definition].

2. Service 366 (26A) Anything other than goods [Introduced vide 101st Constitutional Amendment Act].
3. State 366(26B) With reference to articles 246A, 268, 269, 269A and Article 279A includes a Union territory with Legislature, [Introduced vide 101st Constitutional Amendment Act]

"Goods and Services Tax" law while having unique principles, has significant elements of prior Central and State laws; and is also inspired by VAT/GST legislation of EU, Australia, Malaysia etc., along with International VAT/GST guidelines of OECD.

9.10.2 Registration and Threshold

- Threshold limit of aggregate turnover for exemption from registration and payment of GST for suppliers of services would be ₹ 20 lakhs and ₹ 10 lakhs (for States of Manipur, Mizoram, Nagaland and Tripura).
- Threshold limits of aggregate turnover for exemption from registration and payment of GST for the suppliers of goods would be ₹ 40 lakhs and ₹ 20 lakhs (in the States of Arunachal Pradesh, Manipur, Meghalaya, Mizoram, Nagaland, Puducherry, Sikkim, Telangana, Tripura and Uttarakhand) with effect from 01.04.2019.
- The following classes of taxpayers shall be exempted from obtaining registration: Suppliers of services, having turnover up to ₹ 20 lakh, making inter state supplies. Suppliers of services, having turnover up to ₹ 20 lakh, making supplies through e-commerce platforms.
- Taxpayers may opt for multiple registrations within a State/Union territory in respect of multiple places of business located within the same State/Union territory.
- Mandatory registration is required for only those e-commerce operators who are required to collect tax at source.
- Registration to remain temporarily suspended while cancellation of registration is under process, so that the taxpayer is relieved of continued compliance under the law.
- Revocation of cancellation of those registration, which were cancelled till 31.03.2019, has been allowed. The application for revocation can be filed till 22.07.2019.

9.10.3. Composition Scheme

1. Composition scheme has been formulated for small businessmen being supplier of goods and supplier of restaurant services. Under the scheme, person with annual turnover up to ₹ 1.5 crore (₹ 75 lakhs in States of Arunachal Pradesh, Manipur, Meghalaya, Mizoram, Nagaland, Sikkim, Tripura and Uttarakhand) needs to pay tax equal to 1% to 5% on his turnover and needs to file his returns annually with quarterly payment from FY 2019-20.
2. Composition scheme has been made available for suppliers of services (to those who are not eligible for the presently available Composition

Scheme) with a tax rate of 6% (3% CGST +3% SGST) having an annual turnover in the preceding FY up to ₹ 50 lakhs. They would be liable to file one Annual Return with quarterly payment of taxes. This has been made effective from 01.04.2019.

3. Taxpayers under Composition scheme have been allowed to pay "self assessed tax" on a quarterly basis till 18th of the month succeeding such quarter and furnish a return till 30th April for the previous financial year.
4. A taxpayer who wants to opt for Composition Scheme for a financial year or during the middle of a financial year has to inform the government about his choice by filling FORM GST CMP-02.
5. The GST Council in its 36th meeting held on 27.07.2019 decided that the last date for filing of intimation, in FORM GST CMP-02, for availing the option of payment of tax under notification No. 2/2019-Central Tax (Rate) dated 07.03.2019 (by exclusive supplier of services), be extended from 31.07.2019 to 30.09.2019.
6. The last date for furnishing statement containing the details of the self-assessed tax in FORM GST CMP-08 for the quarter April, 2019 to June, 2019 (by taxpayers under composition scheme), has been extended from 31.07.2019 to 31.08.2019.
7. Composition scheme shall not be available to inter-state suppliers and specified category of manufacturers.

9.11 TAXATION BENEFITS TO SMALL-SCALE INDUSTRY

Various tax benefits are offered by the Government to the small-scale industries in the country.

1. Need for Tax Benefits

Small-scale industries are sensitive due to small resources and capital. In the beginning these industries have to incur more expenses and the return may be nominal. These industries need support to tide over crucid initial stage, to enable them to survive. The Government either exempts them from tax or provides concessional tax liability. This helps small industries accumulate capital and plough back profits in business.

2. Tax Holiday

Under Section 80J of Income Tax Act, 1961, a new small-scale industry is exmepted for the payment of income tax on their profits subject to a maximum of 6% per annum of their capital employed for a period of five year from the commencement of production.

3. Depreciation

Under section 32 of the Income Tax Act, 1961, a small-scale industry is entitled to a deduction on depreciation account on group of assets to the prescribed rate.

4. Rehabilitation Allowance

A rehabilitation allowance is granted to small-scale industries under section

33 of Income Tax Act, 1961 whose business is discontinued due to floods, cyclones, earthquakes, civil riots, accidental fire or war.

5. Investment Allowance

The investment allowance under section 31-A of the Income Tax Act, 1961, is allowed at the rate of 25% of the cost of new plant and machinery installed.

6. Expenditure on scientific research

Under section 35 of Income Tax Act, 1961, expenditure incurred on scientific research is exempted from income tax.

7. Amortisation of Certain Preliminary Expenses

The small-scale industries can write off the preliminary and development expenses incurred in setting up a new enterprise under section 35D of the Income Tax Act, 1961.

8. Backward Areas

A small-scale unit established in backward area under section 80-HH is allowed a deduction of 20% on its profits and gains.

9. Rural Area

Small-scale industries established in rural areas are entitled to a deduction of 20% of the profits and gains for a period of 10 years.

10. Expenditure on Acquisition of Patents and Copyrights

Under section 35-A of the Income Tax Act, 1961, any expenditure of a capital nature incurred in acquiring a patent and copyright by a small-scale industry is deductible from its income.

9.12 LABOUR LAWS

There are a large number of labour legislation/acts. Some important acts have been discussed below:

1. Factory Act, 1948.
2. Payment of Wages Act, 1936.
3. Workman Compensation Act, 1923.
4. Industrial Dispute Act, 1947.
5. Employees State Insurance Act, 1978.
6. Employees Provident Fund Act, 1952.

The labour legislations are prepared to promote:

1. Social Justice.
2. Social Welfare.
3. National Economy.
4. International uniformity.

The need of labour laws is:

1. For improving industrial relations and maintaining industrial peace.
2. To save the workers from exploitation by employers.
3. To avoid industrial disputes.
4. For payment of fair wages to the workers.
5. For payment of compensation to accident victims.

6. For reduction of conflicts and strikes.
7. Job security and compensation during lay-off period.
8. For ensuring good working conditions.
9. For fixing hours of work and rest periods.

9.12.1 Workman Compensation Act, 1923

There is the possibility of more accidents in the evergrowing complexity of industry. The Act provides for cheaper and quicker disposal of disputes relating to compensation. The worker (or his dependents) can claim compensation if the injury has been caused by an accident in the course of employment. The workers are protected against financial hardships arising out of accidents.

The Act also gives the rates of compensation and method for calculation of amount of compensation.

The full details of the Act are enclosed as Annexure–VII.

9.12.2 Payment fo Wages Act, 1936

The objective of the Act is to ensure regular and timely payment of wages to the employed persons, to prevent unauthorised deductions being made from wages and arbitrary fines being imposed on the employed persons.

The full details of the Act are enclosd as Annexure–VIII.

9.12.3 Minimum Wages Act, 1948

Granting of minimum wages to the workers is one of the essential requirements. The wages should be sufficient to maintain the worker and his family. The object of the Act is to prevent exploitation of the workers. The Minimum Wages Act is a piece of social legislation intended to do social justice to workers. By granting minimum wages, industrial peace is maintained.

The salient features of the Act are given in Annexure–IX.

9.12.4 Employees State Insurance Act, 1948

The objective of the act is to provide certain benefits to employees in case of sickness, maternity and employment injury. The Act wants to give such benefits to the employees which are difficult for individual factory owner to provide for his own employees.

The administration of the Act is done by Employee's State Insurance corporation which is an autonomous body.

The various benefits are:

1. Skillness benefit
2. Maternity benefit
3. Disablement benefit
4. Dependent benefit
5. Medical benefit

The employer deducts the contribution of the employees from his salary. After adding his shares, the employer deposits the amount with E.S.I. corporation. The rates of contribution of employee and employer are given in the Act.

The full details of the Act are enclosed as Annexure–X.

9.12.5 Employees' Provident Fund Act, 1952

The Act provides some provisions for the future of industrial worker when he retires and for the benefits of his dependents in case of his death. Retirement benefits include provident fund, family pension and deposit linked insurance. The rate of contribution is 12% by the employee and a matching contribution by the employer.

The salient features of the Act are given in Annexure–XI.

9.12.6 Industrial Dispute Act, 1947

The act provides guide lines for handling and settlement of industrial disputes through adjudication machinery.

The objective of the Act is to make industrial peace through voluntary negotiations and compulsory adjudication.

The full details of the Act are enclosed as Annexure–XII.

9.13 INCENTIVES AND TAX BENEFITS FOR EXPORT

Government has made institutional arrangements to provide export assistance, incentives and tax benefits to boost exports.

9.13.1. Export Incentives

Some of the important incentives are as follows:

1. **Import Replenishment (REP) Licences:** Government has prepared a list of items of imports allowed to exporters. These items (raw materials and components) were used in the manufacture of exported goods.
2. **Import Export Pass Book Scheme** to enable exporters to import duty free raw materials.
3. **Import of Canalized inters against REP licence:** The holder of REP can import items given in the licence.
4. **Import of Capital Goods** against REP licence.
5. **Duty Exception scheme** to import raw materials, components, consumables and spare parts duty free for export products.
6. **100% Export Oriented units** are exempted from import licence formalities and can import capital goods, raw material, components, consumables spare parts under open general licence.
7. **Cash Assistance** as percentage of FOB value of exports to meet competition in foreign markets.
8. **Duty Draw back and Excise Rebate** to give relief of customs and excise duty to experters.
9. **Exception from Excise duty** on finished goods meant for export.

9.13.2. Tax Benefits

Incentives under Income Tax Act are as follows:

1. **Industrial units in Free Trade Zone:** Complete tax holiday for 5 years

falling any time within a period of eight years of commencement of production.

2. **Newly established 100% EDU:** Tax exception for consecuture five years within a period of 8 years of commercial production.
3. **Resident Indian Company executing a contract with a foreign governments:** Tax deduction on 50% of profits and gains.
4. **Resident Indian Company exporting products and receivables in convertible foreign exchange:** Tax deduction on export turnover.
5. **Approved Hotels, Travel Agent and Tour Operators:** Tax deduction on earning in convertible foreign exchange.
6. **Resident Indian Company or Indian Exporting Computer Software:** Tax deduction on earnings in convertible foreign exchange.
7. **Resident Indian Company or Indian promoting sales outside India:** Tax deductions on expenditure incurred.
8. Resident Indian company or Indian receiving dividends, technical know-how fees or technical services or royalty fees. The receivables are excempted for computing taxable income.
9. **Sales Tax Excemption:** Exporter can buy goods for export trade without paying any sale tax.

Question Bank

1. What are the main principles of taxation. Explain the followings:
   (*a*) Income tax
   (*b*) Exercise duty
   (*c*) Sales tax
2. Briefly discuss the objective of laws and statutory measures intended for an entrepreneur.
3. What factors are considered for the selection of business organization?
4. Which are various forms of business ownership? Discuss the salient features of Indian Partnership Act?
5. Briefly describe various Industrial Legislation. What is the scope of Factory Act?
6. What is the objective of labour laws in India? Briefly describe various labour laws.
7. Discuss the salient features of Workman's Compensation Act, 1923.

❑ ❑ ❑

Chapter

10 Entrepreneurial Support System

10.1. INTRODUCTION

Entrepreneurship is a function of the gradations of families and society, socio-culture factors, the environment and effective functioning of the support systems including financial and commercial institutes, research, training extension and consulting services as well as large industrial units interested in developing ancillary industries. Although the individual constitutes the most important element in entrepreneurship, the possibility of the success of an entrepreneur generally gets enhanced by the efficient and effective operation of the support system. The support systems, therefore, have to be clear about their respective roles, relevant activities and consequent expectations from the entrepreneur and they should reinforce values of taking in initiative, creativity, searching for alternatives, trying out innovations, planning orientation and risk-taking coupled with personal responsibility.

The main role of the support system is to release the energy of the entrepreneur for the main task of setting up an enterprise. This can be done if various hurdles in his way are minimised. If the entrepreneur is involved in unnecessary procedural complexities, his energy for developing and responding to the challenge is unproductively used or blocked. If the support system can help in simplifying procedures and in the availability of necessary information, the entrepreneur can concentrate better on challenges.

10.2. SOURCES OF INFORMATION

A new entrepreneur requires information about registration, licencing, finance, marketing and production activities. There are various development agencies at state and national level to assist the entrepreneurs. These multiple agencies can be grouped as follows for the type of information needed.

1. General information.
2. Technical/marketing expertise in specific areas.

3. Technical and financial expertise.
4. Implementation assistance for turn-key projects.

The expert information will help in decision-making process. The classification of needs and agencies in given in Table 10.1. The important agencies are :

1. District Industries Centre (DIC)
2. Directorate/Commissioner of Industries Office
3. State Financial Corporation
4. Technical Consultancy Organisations
5. Agencies Conducting Entrepreneurship Development Programmes

The list also includes:

6. National Laboratories and Institutes
7. Commodity Boards
8. Specialised Trade Publications

The table uses abbreviation for the names of agencies which are explained in Annexure-XIV.

TABLE 10.1. Sources of Information

S.No.	Area	State level agencies	National level agencies
1.	Project selection	SISI, DIC, IDC'S, IIC's, TCO's, SFCs, SIC, IC, IEB, PTC	CB, SIDO, CSIR, DEP, IIC, IFCI, IPB, NRDC, EDI
2.	Registration and Licences	CIF, DDCA, DIC, EB, GMD, SIC, WPCB, IC, LA, STC, TC	CECD, CCIE, ISI, IDC, MID, NSIC, RC, RT, SC, DGTD
3.	Finance	DIC, Bank, SFC, SIC, IICs, IDCs	CB, CEC, ICICI, IDBI, IFCI, NSIC, SBI, DIC
4.	Technical	DDCA, DIC, DJCII, TCOs, GMD	CIPET, CSIR, IIC, IIFT, MRDC, NSIC, RT, SBS, SISI, CITD, ICMR
5.	Training	EDPs, SISI, TCDs, DICs	SBI, CB, CIPET, IRL, NISIET, IITs, NISBUD, EDI
6.	Infrastructure Facilities	DIC, EB, IDC, LA	—
7.	Raw Materials	DIC, MID, MDC, SIC, IC, STC	CCIE, MMTC, MDC, SPC
8.	Plant and Machinery	DIC, IIC, SFC, SEC, IC, IDB	CCIE, NSIC, SISI
9.	Marketing Information	DIC, TCO's, SEC, SIC, RIMCO	DEP, DGSD, CCIE, IIFT, MID, SIC, ICMR, ICAR

Table 10.1 (Contd)...

S.No.	Area	State level agencies	National level agencies
10.	Product standardization, invention, quality marks	DIC, IC, IRL, Technical Research Laboratories, IITs and Technical Institutions	IPB, ISI, RT, SISI, CITD, ICMR

10.3. APPLICATION FORMS

Entrepreneurs are required to fill up application forms during planning, implementation and operational stage of enterprise to meet legal and other requirements. The information is given in Table 10.2.

TABLE 10.2. Application Forms

S.No	Subject	Agencies
(A)	**Planning Stage**	
1.	Provisional registration No.	DIC
2.	Application for shed or plot	SIDC
3.	No objection centrificate from local authorities	LA
4.	No objection certificate from health department	District health officer
5.	No objection certificate from electricity department	Electricity department
6.	Loan applications for term loan	SFC/NB/NSIC
7.	Subsidy registration	DIC
8.	Application for Building Plan and Estimates	Approval of architect/ contractor
9.	Applications for Bank Account/ Cash Credits/Working Capital Loan	NB
10.	Applications for Air and Water Pollution No Objective Certificate	State pollution control authority
11.	Application for the approval of production programme for certain restricted items	DIC, SISI, Central Ministry
12.	Registration of Partnership Deed	Registrar of firms
13.	Application for Ancillary units	Parent companies
14.	Registration of firms	Register of firms
15.	Application for the Boilers and plant layout of the unit	Inspector of Boilers
16.	Application for the production of petroleum based products	Ministry of Petroleum
17.	Application for Excise Registration Number	Excise Department
18.	Application for Latex in rubber based products	Rubber Board
19.	No objection certificate from Forest Department for wood based products	State conservator of forests
20.	Applications for essential commodity items as raw materials	District Civil Supply Department

Table 10.2 (Contd)...

S.No.	Subject	Agencies
21.	Application for imported raw materials	DIC/Export-Import Boards
22.	Application for import of machines	DIC/Export-Import Boards
23.	Application for raw materials quota	D/C/Export-Import Boards
(B)	**During Implementation of Project**	
24.	Application for power connection	Local Electricity Deptt
25.	Application for water	LA
26.	Application for C-Form (Sales Tax)	Sales Tax Department
27.	Application for state Sales Tax Registration	Sales Tax Department
28.	Application for central Sales Tax Registration	Sales Tax Department
29.	Application for exemptions from Sales Tax	DIC/Sales Tax Deptt.
30.	Application for exemption from Octroi Duty.	DIC/LA
31.	Application for storing of inflammable raw material	Director of Explosies
(C)	**During Running of Enterprise**	
32.	Application for Permanent Registration Number	DIC/Directorate of Industries
33.	Application for Subsidy Claims	DIC
34.	Application for Power Subsidy	LA
35.	Application for Food Preservation Ordinance Licence	Food Controller
36.	Application for registration in case of more than 20 employees without power use or more than 10 employees with power use.	Labour Welfare Board/ Employment Exchange/ P.F. Commissioner
37.	Application for Product Marketing to the Central Government Departments	DIC, DGSD

10.4. DISTRICT INDUSTRY CENTRES (DICS)

District Industry Centres have been established in each district to provide all services and facilities to small and village industries under one roof. Its main responsibility is to act as chief coordinator or multifunctional agency of various government departments and other agencies.

Organisational setup

Each DIC has one General Manager and seven managers to look after the following functions separately.

The success of the centre will depend upon the effective leadership of the General Manager or proper coordination among managers.

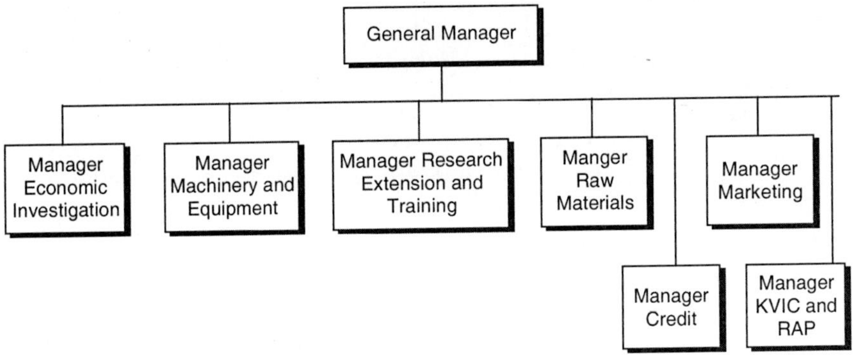

Fig. 10.1.

Functions

District Industries Centres provide pre-investment, investment and post-investment assistance to entrepreneurs under one roof. These centres contribute in promotion of small industries, development of entrepreneurship and generation of self-employment. These centres have to play a leading role in district level industrial development through the following functions:

1. Identification of Entrepreneurs

DIC conducts entrepreneurial motivation programmes throughout district and develops new entrepreneurs.

2. Selection of Projects

DIC gives technical guidance to new entrepreneurs for the selection of suitable projects.

3. Provisional Registration under SSI

Entrepreneurs are issued provisional SSI registration after selection of projects. This registration is essential for getting assistance from financial institutions.

4. Purchase of Fixed Assests

DIC sponsors loan applications to TIIC, SIDCO and banks for purchase of land and buildings, purchase of plant and machinery. It also sanctions margin money under Rural Industries Project Loan Scheme payable to other financial agencies.

5. Clearances from Various Departments

It issues No Objection Certificates and takes initiative to get clearance from various departments. It takes follow up actions for power connection.

6. Assistance to Raw Material Supplies

It makes necessary recommendations to the concerned raw materials suppliers and issues the required certificates for the import of raw materials and machinery where required.

7. Assistance to village Artisans and Handicrafts

DIC arranges the financial assistance from the lead bank or nationalised banks in the respective areas.

8. Interest-free Sales Tax Loan

DIC issues sanction order for interest-free Sales Tax Loan available for SIDCO. This loan is available to SSI units, set up in rural areas. The maximum limit of loan is 8% of the total fixed assets.

DIC also recommends the SSI units to National Small Industries Corporation (NSIC) for registration for Government Purchase Programme.

9. Subsidy Schemes

DIC assists SSI units and rural artisans (tiny industries) to get subsidies on power, interest for engineers and IRDP Scheme from concerned institutions.

10. Training Programmes

DIC conducts training programmes for rural entrepreneurs. It also assists other agencies involved in training of small entrepreneurs.

11. Self-employment for Unemployed Educated Youth

This scheme was introduced in 1983-84 for youths with school certificate between the age of 18 to 25 years. Technocrats and women get preference.

10.5. COMMERCIAL BANKS

Many commercial banks are rendering two types of help to the entrepreneurs.

1. Extension of Finance

Bank credit is the major source of finance for working capital. Commercial banks offer both secured as well as unsecured loans to entrepreneurs such as cash credit, overdrafts, loans and advances and purchase and discounting of bills. The bank provide credit against security.

A loan may be secured by tangible assets or by person security either by lien, fledge, hypothecation, mortgage or charge, etc. Bank loans are an important source of enterprise financing such as seasonal build ups in accounts receivable and inventories.

Banks do not provide 100% finance. They insist that the customers should bring a portion of finance from other sources.

2. Entrepreneurship Development Programmes (EDPs)

The commercial banks are promoting new small scale entrepreneurs through EDPs. They are holding EDPs in collaboration with specialised institutions such as DIC, SISI, TCOs etc. EDPs are conducted to identify entrepreneurs specially in backward areas, train and monitor them to start new ventures.

A commercial bank acts as a lead bank for a particular district and render the above help.

10.5.1. State Bank of India (SBI)

Bank's EDP consists of three phases:

(*i*) Entrepreneurship Awareness Camps for entrepreneurial opportunities.

(*ii*) Training programmes in developing motivation and managerial skills.

(*iii*) Counselling, encouragement and infra-structural support for setting up and running an enterprise.

The bank has launched the following special schemes :

1. Package of financial arrangement and assistance to small-scale units for promotional and expansion activities and issue capital fund.
2. Research and Development Fund of Rs. 5 crore for assissting entrepreneurial development.
3. **Equity Fund Scheme:** The bank may grant interest-free loan upto 25% of project cost which is the minimum contribution of an entrepreneur.
4. **Techno-entrepreneurs:** The bank extends financial assistance upto 100% if necessary. The target group is technocrats and trained entrepreneurs, who lack financial capability to meet the normal margins stipulated by the Bank.

10.5.2. Punjab National Bank (PNB)

The bank offers the following package of assistance through its Merchant Banking Division as a leading bank.

1. Help in the preparation of market-survey report with the assistance of technical consultant.
2. Study of economic viability and technical feasibility of the proposals.
3. After receipt of application for letter of intent the bank provides help to entrepreneurs in obtaining various government consents required for industrial projects.
4. The bank assists the entrepreneur in raising capital resources from financial institution in the form of debentures, term loans, deffered payment guarantees, etc.
5. The bank assists in raising foreign exchange resources required for import of plant and machinery, components, raw materials, etc. Arrangement is made through the Bank's foreign correspondents, supplier's credit, buyer's credit and foreign currency loans.
6. The bank determines the capital structure, assists in obtaining consent of SEBI, finalization of syndication of underwriting arrangements, handling of share applications and relative allotment in consultation with stock exchange, etc., financing export of capital equipment on deferred payment loans.
7. The bank gives suggestion in strengthening the capital base of small scale industries which intend to expand/diversity by conversion of partnership firm into private limited companies or conversion of private limited companies into public limited companies.

10.5.3. Indian Bank

The Entrepreneurship Service cell of the bank arranges a meeting of appropriate panel member with entrepreneur. The cell and the member assist the entrepreneur. This service is available only at Madras and a few selected centres. The areas of assistance are :

1. Consultancy service from product identification to its implementation.
2. Marketing through the bank personnel and panel of expert specialists.
3. Consultancy services to graduates and technical graduates, unemployed engineers and diploma-holders and business executives.

10.5.4. Bank of Baroda

(*a*) Entrepreneurial Banking Service : The bank has started Entrepreneurial Bank Service in collaboration with Uttar Pradesh Small Industries Corporation to assist technician entrepreneurs to set up their own units at Rai-Bareli Industrial complex. The scheme includes :

1. Building and hardware materials
2. Acquisition of fixed assets
3. In-plant training in established industrial units
4. Working capital facilities.

(*b*) Multi-Service Agency at Mumbai : The bank provides technical assistance to self-employed persons for :

1. Feasibility study of project proposals.
2. Guidance to availability of raw materials, etc.
3. Guidance for marketing prospects.

10.5.5. Bank of India

The bank has set up an Entrepreneurial clinic-cum-guidance service to foster growth of entrepreneur and economic development. The scheme offers :

1. Assistance in selection of industry, preparation and evaluation of project report and market survey.
2. Practical training in the line, if necessary.
3. Assistance in obtaining government clearance, procurement of machinery and equipments and marketing of products.
4. Assistance and guidance in implementation of project.

The novel feature of the scheme is that the bank will provide from a panel of industrialists a 'foster father' to guide and assist the budding entrepreneurs.

10.5.6. Canara Bank

The bank has set up Industrial Information and Guidance Service for the following :

1. To provide information and advise to entrepreneurs for scope for establishment of industries
2. Technical and marketing facilities

3. Taxation
4. Export and imports
5. Accounting and management
6. Preparation of project report on preferred industries

10.5.7. Grindlays Bank Limited

(*a*) Small-Scale Consultancy Service, Kolkata

It offers assistance in :

1. Preparing project feasibility reports.
2. Conducting overall industry studies.
3. Marketing and sales.
4. Management accounting.
5. Detailed in-depth study of existing systems and procedures in various areas of operation for problems and other deficiencies, and develop tailor-made solutions.
6. Help to management in the implementation of above solutions.

(*b*) Merchant Banking Division, Mumbai, Chennai, Kolkata, New Delhi. This division helps in :

1. Managing public issues for raising capital
2. Establishing liaison with government to lending bodes for raising capital.
3. Development of term finance.
4. Management audits to assess strengths and weakness of the enterprise.
5. Tailor-made solutions to eliminate the weaknesses.

10.5.8. Syndicate Bank

The bank has established a technical cell of engineers in Small-Scale Industries Deptt. The cell is working at Manipal and different industrial growth centres and offers following services :

1. Approval of viability of credit proposals from small scale industries sector.
2. Counselling to entrepreneurs including help in the preparation of project reports.

10.5.9. Other Banks

The following commercial banks have also special entrepreneurship development services.

1. United Bank of India : Technical cell at Kolkata.
2. Union Bank of India : General Services cell at Mumbai.
3. United Commercial Bank : Cell at Kolkata.
4. Indian Overseas Bank : Small Business Aid Agency at Chennai and Mumbai.

10.5.10. New Entrepreneur Development Agency, 1986

The New Entrepreneur Development Agency has been created to :

1. Help educated unemployed graduates to set up SSI units in urban and metropolitan areas.
2. Selection of candidates only from among those sponsored by Government Institutions and Agencies like SISI, DIC's, Employment Guidance Bureau, etc.
3. The selected candidates are assisted in selection of projects suitable to their aptitude and backgrounds.
4. Entrepreneurship Development Programme for selected candidates by SISI, TCO's.
5. Practical training is arranged in industries.
6. Marketing tie-up with user industries after registration with Directorate of Industries and Commence or DICs.
7. Credit requirements upto Rs. 5 lakhs are sanctioned.

10.6. SMALL INDUSTRIES SERVICE INSTITUTES (SISIs)

The Small Industries Service Institutes have been set up in state capitals and other places all over the country to provide consultancy and training to small entrepreneurs both existing and prospective.

The main function of SISIs include :

1. To serve as interface between Central and State Governments
2. To render technical support services.
3. To conduct Entrepreneurship Development Programmes.
4. To initiate promotional programmes.

The SISIs also render assistance in the following areas :

1. Economic Consultancy/Information/EDP Consultancy.
2. Trade and Market information.
3. Project profiles.
4. State industrial potential survey.
5. District industrial potential surveys.
6. Modernisation and inplant studies.
7. Workshop facilities.
8. Training in various trade/activities.

10.7. NATIONAL SMALL INDUSTRIES CORPORATION (NSIC)

The National Small Industries Corporation Ltd. (NSIC), an enterprise under the Union Ministry of Industries, was set up in New Delhi in 1955 to promote, aid and facilitate the growth of small scale industries in the country. NSIC offers a package of assistance for the benefit of Small Enterprises.

1. Single Point Registration

Registration under this scheme for participating in Government and Public Sector undertaking tenders.

2. Marketing Assistance Programme

NSIC participates in Government tenders on behalf of small enterprises to procure orders for them.

3. Information Service

NSIC gets continuously updated with the latest specific information on business leads, tenders, technology and policy issues.

4. Performance and Credit Rating

NSIC gives Credit Rating by international agencies subsidized for small enterprises upto 75% to get better credit terms from banks and export orders from foreign buyers.

5. Raw Material Assistance

NSIC fulfils raw material requirement of small enterprises and provides raw material on convenient and flexible terms.

6. Meeting Credit Needs of Small Enterprises

NSIC facilitates sanction of term loan and working capital credit limit of small enterprise from the banks.

10.8. SMALL INDUSTRIES DEVELOPMENT BANK OF INDIA (SIDBI)

Government of India has set up Small Industries Development Bank of India (SIDBI) in 1989 as a wholly-owned subsidiary of Industrial Development Bank of India (IDBI) to ensure larger flow of financial and non-financial assistance to small-scale sector. It has its head office in Lucknow.

The important functions of SIDBI are as follows :

1. To initiate steps for technological upgradation and modernisation of existing units.
2. To expand the channels for marketing the products of SSI sector in domestic and international markets.
3. To promote employment oriented industries specially in semi-urban areas to create more employment opportunities and thereby checking migration of people to urban areas.

Special Programmes of SIDBI

1. Financial assistance to small-scale industries is channelised through existing credit delivery system comprising State Financial Corporations, State Industrial Development Corporations, Commercial Banks and Regional Rural Banks.
2. Equipment finance scheme for providing direct finance to existing well-run small-scale units taking up technology upgradation/modernisation.
3. Venture capital fund of Rs. 10 crore was launched exclusively for small-scale units.

4. SIDBI provides financial support to NSIC for providing leasing, hire-purchase and marketing support to the industrial units in the small-scale sector.

10.9. IMPORTANT INSTITUTIONS AT STATE LEVEL

Various support institutions have been set up to assist the small entrepreneur at the state level.

1. Small Industries Development Organisation (SIDO)

SIDO is a policy-making, coordinating and monitoring agency for the development of small scale entrepreneurs. It maintains a close liaison with government, financial institutions and other agencies involved in the promotion and development of small scale units. It provides a comprehensive range of consultancy services and technical, managerial, economic and marketing assistance to SSI units.

2. Directorate of Industries of the State Government

Directorate of Industries works for development of industries in general and small-scale industries in particular. It is staffed with administrative and technical officers. It runs various training schemes, production schemes and common facilities schemes. It also provides facilities of developed industrial land and factory sheds in industrial states, allocate quotas of scarce raw materials, certify import requirements and organise industrial cooperatives.

3. Small Industries Corporation (SIC)

SIC distributes scarce raw materials, supplies machinery on hire-purchase basis, constitutes and manages industrial estates processes orders from Government Departments, assists in export marketing and provides financial, technical and managerial assistance to small enterprises in specific areas.

4. Small Industries Development Corporations (SIDCs)

It plays a lead role in developing small scale sector. It provides constructed sheds/plots in industrial state, processes scarce key raw materials, financial assistance and marketing assistance to small entrepreneurs.

5. State Financial Corporations (SFCs)

State Financial Corporation assists the medium and small scale industries by providing financial assistance in the form of term loan alone or in supplementing the financial assistance from national financial corporations and commercial banks. It grants loans and subscribes to the debentures, stocks, bonds. It also assists under seed capital/soft loan scheme. It also assumes the role of promotional agency in addition to financing institution.

6. Technical Consultancy Organisation (TCO)

Technical Consultancy Organisations have been set up in each state by All India Financial Institutions in collaboration with state-level financial/ development institutes and commercial banks to cater to the consultancy needs of small industries and new entrepreneurs. Its functions include preparation of project proposal and feasibility profiles, industrial surveys, identifying potential entrepreneurs, supervision of projects, export consultancy, conduct of EDPs, merchant banking service.

10.10. NATIONAL BANK FOR AGRICULTURE AND RURAL DEVELOPMENT (NABARD)

1. Objectives

The rural financial system in the country calls for a strong and efficient credit delivery system, capable of taking care of the expanding and diverse credit needs of agriculture and rural development. More than 50% of the rural credit is disbursed by the Co-operative Banks and Regional Rural Banks. NABARD is responsible for regulating and supervising the functions of Co-operative banks and RRBs. In this direction NABARD has been taking various initiatives in association with Government of India and RBI to improve the health of Co-operative banks and Regional Rural Banks.

2. Introduction

1. NABARD is an apex institution accredited with all matters concerning policy, planning and operations in the field of credit for agriculture and other economic activities in rural areas.
2. It is an apex refinancing agency for the institutions providing investment and production credit for promoting the various developmental activities in rural areas.
3. It takes measures towards institution building for improving absorptive capacity of the credit delivery system, including monitoring, formulation of rehabilitation schemes, restructuring of credit institutions, training of personnel, etc.
4. It co-ordinates the rural financing activities of all the institutions engaged in developmental work at the field level and maintains liaison with Government of India, State Governments, Reserve Bank of India and other national level institutions concerned with policy formulation.
5. It prepares, on annual basis, rural credit plans for all districts in the country; these plans form the base for annual credit plans of all rural financial institutions.
6. It undertakes monitoring and evaluation of projects refinanced by it.
7. It promotes research in the fields of rural banking, agriculture and rural development.

3. Supervisory Role of NABARD

NABARD has been entrusted with the statutory responsibility of conducting inspections of State Co-operative Banks (SCBs), District Central Co-operative Banks (DCCBs) and Regional Rural Banks (RRBs) under the provision of the Banking Regulation Act, 1949. In addition, NABARD has also been conducting periodic inspections of state level co-operative institutions such as State Co-operative Agriculture and Rural Development Banks (SCARDBs), Apex Weavers Societies, Marketing Federations, etc. on a voluntary basis. Although the prime objective of the statutory inspections is to ensure general safety of public deposits,

NABARD, through these statutory inspections, has been simultaneously endeavouring for further developing and strengthening the above institutions to enable them to play a far more effective and efficient role in meeting the rural credit requirements.

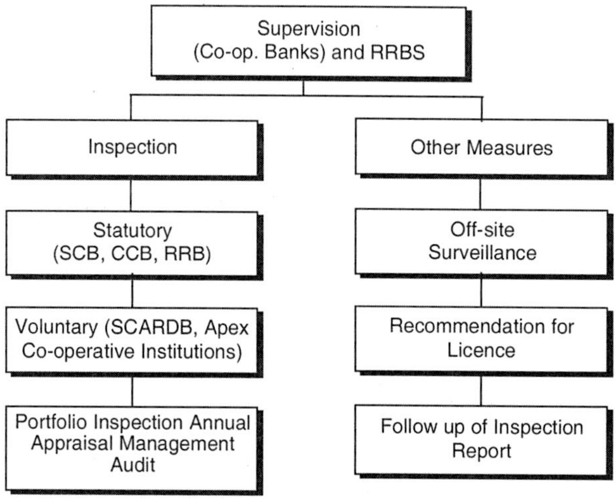

Fig. 10.2.

The general banking environment emerging out of the financial sector reforms introduced by GOI/RBI has also since been extended to cover co-operative banks and RRBs. While the capital adequacy norm has not yet been made applicable to these banks, the other prudential norms viz., income recognition, asset classification and provisioning, which were made applicable by Reserve Bank of India to the commercial banking sector have been extended to cover SCBs and DCCBs since 1996-97 and to SCARDs in 1997-98. These norms had already been extended to RRBs since 1995-96. The exposure of these institutions to the prudential norms also called for a suitable strategy to be adopted by the NABARD to help these banks to adjust to the new financial discipline.

The changes mentioned above necessitated a review and revision of the adopted by the NABARD for the inspection and supervision of the cooperative banks and RRBs. Under the revised strategy, a sharper focus of the NABARD's inspection was given on the core areas of the functioning of banks pertaining to Capital Adequacy, Asset Quality, Management, Earnings, Liquidity and Systems Compliance (CAMELSC). Thus NABARD's focus in its statutory 'on-site' inspections is on core assessments leaving the collateral appraisals to supplementary inspections. The micro level aspects are to be taken care of by the banks themselves by way of internal inspections or by other agencies such as auditors. In this direction, through a series of workshops and meetings held with the Chief Executives and the Chief Auditors of cooperative banks, NABARD attempted to ensure that the other areas, particularly relating to the internal checks and controls, revenue and income realisation by way of interest on loans and deposits and other routine features of carrying out general banking transactions were suitably taken care of by the respective banks and their concurrent/statutory audit systems.

4. Off-site Surveillance

As a part of the new strategy of supervision, a system of 'Off-site Surveillance' has been introduced as a supplementary tool to the on-site inspection, its objectives are to obtain and analyse critical data on a continuous basis, to identify areas of supervisory concern and to identify early warning signals and risky areas requiring further probe. The system basically envisages desk scrutiny of operations of cooperative banks and RRBs through a set of statutory and non-statutory returns. While the periodical statutory on-site inspections attempt an overall evaluation of the performance of the banks with a stipulated period, off-site surveillance envisages continuous supervision supplementing the on-site inspections with additional instruments of supervision.

5. Board of Supervision (for SCBs, DCCBs and RRBs)

Has been constituted by NABARD under Section 13(3) of NABARD Act, 1981 as an Internal Committee to the Board of Directors of NABARD. The broad powers and functions of the Board of Supervision are : giving directions and guidance in respect of policies and on matters relating to supervision and inspection, reviewing the inspection findings, suggesting appropriate measures, reviewing the follow-up action taken by Department of Supervision (DoS) on matters of frauds and internal checks and control, identifying the emerging supervisory issues in the functioning of cooperative banks/RRBs such as NPAs recovery, investment portfolio, credit monitoring system, management practices, frauds, etc., and suggesting necessary follow-up measures. It can also recommend appropriate training for Inspecting Officers of NABARD for imparting necessary skills and knowledge, suggest measures for strengthening of DoS, recommend issue of directions by RBI, oversee the quality of inspections carried out and the reports issued, review the information generated through off-site surveillance and other supplementary vehicles, action taken thereon, and undertake any other functions entrusted from time to time by the Board of Directors of NABARD.

The Board of Supervision, since its formation on 20 November, 1999, has held 22 meetings till 30 June 2004 and reviewed the financial position of Cooperative Banks and RRBs. Based on the observations of BoS, authorities concerned have been apprised of the weaknesses.

10.11 SPECIAL FINANCIAL SERVICES PROVIDED BY INSTITUTES

The special financial services provided by the institutes are :

1. Term lending
2. Underwriting of capital issues.

10.11.1 Term Lending

A firm obtains long-term debt by raising term loans from financial institutions and banks. Term loan is generally repayable in less than 10 years. These are mainly used to purchase fixed assets and working capital margin.

1. Special Features of Term Loans

(i) Currency

Rupee term loans are directly given to the firm for setting up new projects as well as for expansion, modernization and renovation projects.

Foreign currency term loans are given to meet foreign currency expenditure to import plant and machinery and foreign technical know-how fees.

(ii) Security

The term loans represent secured borrowing. The assets financed by the loan provide the prime security and other assets of the company as collateral security.

(iii) Interest and Principal Repayment

The principal amount of a term loan is repayable over a period of 5 to 10 years after an initial grace of 1 to 2 years. The payments are made in equal semi-annual installments. The principal repayment remains constant but interest burden declines over time.

(iv) Restrictive Corenants

Restrictive conditions are improved on the borrower to protect the interest of the institutes. The institutes have a say in the appointment of directors, key personnel, management functions affecting the financial health of the company.

2. Advantages of Term Lending

1. Interest on loan is a tax-deducible expense.
2. The control of the company is not diluted as the financial institutions do not have voting rights.
3. Financial institutions receive only interest and principal and not profits.
4. The maturity of loan can be tailored to suit the company.
5. The cost of raising term loan is very low.
6. The loans are protected against high unanticipated inflation.
7. The loan has disciplining effect on the management of the firm.

3. Disadvantages of Term Lending

1. Defaulting in repayment of principal and interest can cause embarrassment and even bankruptcy.
2. The restriction on borrowing limit of the firm can affect financial and operational flexibility.
3. The real cost of debit will be greater than expected if inflation is unexpectedly low.

10.11.2 Underwriting of Capital Issues

Funds needed for fixed capital are generally raised as follows :

1. Issue of shares of different types and debentures.

2. Ploughing back a part of profit made by enterprise.
3. Long-term deposits from the public.
4. Long-term loans from financial institutions.

1. Share and Debentures

Industries generally raise a major portion of their funds to meet their capital demands by selling shares and debentures. It is easy to raise funds by selling equity shares because there is substantial money with the public.

2. Public Subscription to Capital

Equity capital, after deducting the amount to be statutorily subscribed by the promoters, may be offered to the public for subscription in cash. For a project having a debt-equity ratio of 2 : 1 and promoter's contribution of 20% the promoters would be required to subscribe to 60% of equity capital and balance 40% would be issued to public for subscription. For becoming eligible to listing on a recognized stock exchange, companies are required to issue 60% of their capital to public for subscription through a prospectus.

3. New Industry

A new industry promoted by relatively unknown entrepreneurs have the limitation of attracting public subscription to capital. Therefore, they go to financial institutions for underwriting their public issues for raising capital for certain time till they are able to attract the confidence of the public through their performance.

4. Underwriting of Capital Issue

All leading financial institutions at national and state level are underwriters of capital issue.

Underwriting the issue of stock, shares, bonds or debentures by industries is provided by the institutions on certain conditions :

(*i*) The industry does not retain any shares.

(*ii*) The underwriting liabilities are not beyond a period of 7 years.

(*iii*) The underwriting fees may be 2.5% of the capital.

10.12 INDUSTRIAL FINANCE CORPORATION OF INDIA (IFCI)

10.12.1 Incorporation and Purpose

IFCI was established in 1948 under the Act of Parliament with the objective of providing medium and long term credit to industrial enterprizes in India. It was transferred into corporation in 1993 to provide greater flexibility in order to meet the needs of changing financial system.

10.12.2 Management

The management consists of a whole-time chairman appointed by the Central Government after consultation with IDBI. It has twelve directors appointed by Central Government (2), IDBI (4) and six directors are elected by shareholders.

10.12.3 Project Finance

IFCI funds projects costing above Rs. 300 lakh in the following forms :

(i) Rupee and foreign currency term loans.
(ii) Undertaking of share and debenture issues.
(iii) Direct subscription to equity
(iv) Guarantees
(v) Soft loans
(vi) Equipment financing

10.12.4 Types of Activities

IFCI is authorized to undertake following types of activities.

1. Loans repayable within 25 years.
2. Underwriting the issue of stock, shares, bonds or debentures by industries with liabilities beyond 7 years.
3. Guaranteeing loans, deferred payments, debentures raised by industrial undertaking from different sources.
4. Subscribing to stock or shares of any industrial concern.
5. Agent for Central Government for loans or debentures subscribed by International Bank for Reconstruction and Development (IBRD).

10.12.5 Functions and Policies

The following organizations can avail of financial assistance.

1. Limited company or co-operative society engaged in manufacture, preservation of goods.
2. Shipping industry.
3. Mining industry.
4. Hotel industry.
5. Power industry for generation and distribution.
6. Public sector projects.
7. New industrial projects by Central Government in less developed districts in States and Union Territories.

10.12.6 Sources of Funds

Borrowed from the market by issue of bonds, loans for the Central Government and foreign credits.

10.12.7 New Promotional Schemes

In order to encourage new entrepreneurs and technologies upgradation, following promotional schemes have been framed.

1. Consultancy Fee Subsidy Scheme

It is a scheme of subsidy for consultancy to industries in specific areas.

2. Interest Subsidy Scheme

Scheme of interest subsidy for self-employment of young unemployed persons, women entrepreneurs, quality control measures in small scale industries, adoption of indigeneous technology, etc.

3. Entrepreneurship Development Schemes

Scheme for encouraging entrepreneurship development in tourism and related activities, persons rendered jobless, revival of sick units.

10.13 THE INDUSTRIAL DEVELOPMENT BANK OF INDIA (IDBI)

10.13.1 Incorporation and Purpose

It was established in 1964 under Industrial Development Bank of India Act. In 1976, the ownership has been transferred to Central Government.

It is the principal financial institution for coordinating the activities of institutions engaged in the field of financing, developing or promoting industry. It has been assigned a special role to play for industrial development.

10.13.2 Objectives and Functions

1. To serve as an apex institution for term finance for industry, coordination of the working of institutions engaged in financing, promoting and developing industries and to assist the development of these institutions.
2. To plan, promote and develop industries to fill the gap in the industrial structure of India.
3. To provide technical and administrative assistance for promotion, management or expansion of industry.
4. To undertake market and investment research and surveys and also technical and economic studies for the development of industry.
5. To finance industries engaged in manufacturing, processing or preservation of goods, or in mining, shipping, transport, hotel industries or in the generation or distribution of power, in fishing maintenance repair, testing or servicing of machinery, vehicles, vessels, for setting up of industrial estates.
6. To help industries engaged in research and development providing special or technical knowledge and other services for the promotion and growth of industries.
7. To provide finance or export of engineering goods or services on deffered payment basis.

10.13.3 Special Schemes

IDBI has following schemes for the benefit of enterprises and entrepreneurs in the small and medium scale sector.

1. Project Finance Scheme

Loans, underwriting, direct subscription and guarantees for

— Modernization Assistance Scheme for all industries.
— Textile Modernization Fund Scheme.
— Technical Development Fund Scheme.
— Venture Capital Fund Scheme.
— Energy Audit Subsidy Scheme.
— Equipment Finance for Energy Conservation Scheme.
— Foreign Currency Assistance Scheme.

2. Refinance Schemes

— Refinance scheme for industrial loans for small and medium industries.
— Refinance scheme for modernization and rehabilitation of small and medium industries.
— Equipment refinance scheme.
— Bills discounting/rediscounting scheme.
— Seed capital scheme.
— Scheme for concessional finance for development of no-industry distincts and other backward areas.
— Scheme of concessional assistance for manufacture and industrialization of renewable energy systems.
— Scheme for investment in shares and bonds of other financial institutions.

3. Soft Loan Scheme

IDBI extends soft loans to units in selected industry groups. It is operated in participation with IFCI and ICICI. The soft loans are extended on concessional terms regarding interest, promoter's contribution, debt-equity ratios, initial moratonium and repayment period.

10.13.4 Sources of Funds

1. Capital contribution from Government.
2. Loan capital from Government.
3. Loan capital from Reserve Bank of India.
4. Bonds from domestic market.
5. Borrowing in foreign exchange from international capital market.
6. Deposits under Investment Deposit Account Schemes.
7. IDBI Capital bond scheme.

10.14 STATE FINANCIAL CORPORATIONS (SFCs)

10.14.1 Incorporation and Purpose

State Financial Corporations operating at state level form an integral part of development financial system in the country. The function and objective of SFCs are financing and promoting small and medium industries for achieving balanced regional socio-economic growth, catalysing higher investment, generating greater employment opportunities and widening the ownership base of industry.

SFCs were set up in 1951 under SFCs Act, 1951.

10.14.2 Special Schemes

— Financial assistance to industrial units in the form of term loans, direct subscription to equity/debentures, guarantees, discounting of bills of exchange.

— Refinancing and equity assistance formution by IDBI/SIDBI.

— Schemes for artisans, SC/ST, women, ex-servicemen, physically handicapped.

— Schemes for transport operators, hotels, tourism, hospitals and nursing homes.

10.14.3 Objectives and Functions

— Term loan for acquisition of land, building, plant and machinery, pre-operative and other assets.

— Promotion of self-employment.

— Encourage new and technically/professionally qualified women entrepreneurs to set up industries.

— Financing of expansion, modernization and upgradation of technology in existing units.

— Financial assistance for rehabilitation of sick units.

— Promotion and expansion of industry by rural and urban artisans.

10.14.4 Package of Assistance

Package schemes of incentives to motivate and encourage entrepreneurs. SFCs provide two types of services:

— Package of assistance for development of entrepreneurs and industries.

— Financial assistance to eligible concerns.

10.14.5 Eligible Concerns for Financial Assistance

The concerns engaged in the following activities are eligible for financial assistance from SFCs.

1. Manufacture of goods.
2. Preservation of goods.
3. Processing of goods.

4. Mining.
5. Generation and distribution of electricity.
6. Hotel industry.
7. Transport of passenger or goods by road or water or air.
8. Maintenance, repair, testing or servicing of machinery, vehicles, vessels, motor boats, trawlers, tractors.
9. Assembly, repair or packing of goods with the aid of machines and power.
10. Development of land as industrial estate.
11. Fishing and shore facilities for fishing.
12. Providing special or technical knowledge or other services for the promotion and growth of industries.

10.15 UNIT TRUST OF INDIA (UTI)

10.15.1 Incorporation and Purpose

UTI was established in 1964 under an Act of Parliament to mobilize savings of small investors through sale of units and channelise them into corporate investment.

10.15.2 Sources of Funds

UTI mobilizes savings of small investors through schemes and plans for common investors. The various schemes are :

1. Grihlakshmi Unit Plan.
2. Retirement Benefit Plan.
3. Primary Equity Fund.
4. Columbus India Fund.
5. Unit Scheme, 1995 targetted at Corporate investors.

10.15.3 Assistance Schemes

UTI provides assistance to corporate sector in the following modes.

1. Term loans.
2. Underwriting of shares and debentures.
3. Direct subscription to shares and debentures.
4. Financial assistance to both private sector and public sector.
5. Financing of new industrial projects.
6. Expansion, diversification, modernization, balancing equipment and rehabilitation of sick units.
7. Working capital loans.

Question Bank

1. What are District Industries Centres (DICs)? Explain the function of the DICs.
2. Discuss the need for institutional support to small-scale industries.
3. What are the functions performed by commercial banks to boost the growth of small-scale industries in the country?
4. What preliminary information is needed by a new entrepreneur? List the sources of information.
5. What type of application forms are required at planning stage, during implementation of project and during running of an enterprise? List the contact agencies.
6. Write short notes on the following :
   (*a*) New Entrepreneur Development Agency
   (*b*) Small Industries Service Institute
7. Discuss the support provided by the National Small Industries Corporation Ltd. (NSIC) to small-scale industries in India.
8. Discuss the role played by the following state-level agencies in the growth of small scale industries.
   (*a*) SIDO
   (*b*) SFC
   (*c*) TCO
   (*d*) SIDC
   (*e*) SIC
9. Give an account of functions performed by National Bank for Agriculture and Rural Development.
10. What is the role of small Industries Development Bank of India (SIDBI) in financing small industrial sector? Discuss the special programmes launched by SIDBI.
11. Discuss in detail the role of following financial agencies in the development of industries in India :
    (*a*) State Financial Corporations
    (*b*) Industrial Finance Corporation of India
    (*c*) Unit Trust of India
12. Explain in brief the contribution of IDBI in the development of industries in India.
13. What is the role of financial institutions in industrial development? List important financial agencies at national and state levels.

❑❑❑

ANNEXURES

Annexure-I	:	Factory Act, 1948
Annexure-II	:	Indian Boiler Act, 1923
Annexure-III	:	Electricity Safety Act, 1910
Annexure-IV	:	The Water (Prevention and Control of Pollution) Act, 1974
Annexure-V	:	The Air (Prevention and Control of Pollution) Act, 1981
Annexure-VI	:	The Indian Partnership Act, 1932
Annexure-VII	:	Workman's Compensation Act, 1923
Annexure-VIII	:	Payment of Wages Act, 1936
Annexure-IX	:	Minimum Wages Act, 1948
Annexure-X	:	Employee's State Insurance Act, 1948
Annexure-XI	:	Employee's Provident Fund Act, 1952
Annexure-XII	:	Industrial Dispute Act, 1947
Annexure-XIII	:	List of Industries Reserved for the Public Sector
Annexure-XIV	:	List of Agencies

Annexure—I

FACTORY ACT, 1948

The working conditions in a factory should ensure health and safety of the workers and other employees. The Factories Act, 1948 is a comprehensive document covering various aspects of design and operation of factories and regulates the working conditions. The Act extends to all parts of India.

1. Salient Provisions

Factory Act, 1948 covers the following

(*i*) Licensing and Registration
(*ii*) Health
(*iii*) Safety
(*iv*) Welfare
(*v*) Hours of work
(*vi*) Employment of young persons
(*vii*) Leave

2. Some Definitions

(*i*) *Adult* is a person above eighteen years of age.
(*ii*) *Adolescent* is a person of fifteen years but less than eighteen years of age.
(*iii*) *Child* is a person below the age of fifteen.
(*iv*) *Young person* under the age of eighteen years. He may be a child or adolescent.
(*v*) *Day* means a period of twenty four hours beginning at midnight.
(*vi*) *Week* is a period of seven days beginning at midnight on any night approved by the Chief Inspector of Factories.
(*vii*) *Power* is electrical energy or mechanical energy but not generated by human or animal power.
(*viii*) *Prime Mover* is a machine to provide or generate power.
(*ix*) *Machinery* includes prime movers, transmission machinery or driven machinery.
(*x*) *Manufacturing Process* is any process for

— making, altering, repairing, ornamenting finishing, packing, oiling, washing, cleaning, breaking up, demolishing or treating or adopting any article or substance for use, sale, transport, delivery or disposal.

— pumping oil, water, sewage or any other substance,

— generating, transforming or transmitting power,
— composing, printing by press, lithography or photograving or other similar process.
— counstructing, recounstructing, repairing, refittering, finishing, breaking up of ships or vessels,
— preserving or storing any articles in cold storage.

(*xi*) *Worker* is a person employed directly or through an agency for wages or otherwise for any manufacturing process or cleaning the machinery or premises or any other work connected with the manufacturing process.

(*xii*) *Factory* is any premises

(*a*) Where ten or more workers are working or were working on any day during last twelve months and manufacturing is done with the aid of power.

(*b*) Where twenty or more workers are working or were working on any day during last twelve months and manufacturing is done without the aid of power.

(*xiii*) **Occupier of factory** is a person who has ultimate control over the affairs of the factory or where the above affairs are entrusted to a managing agent. The managing agent shall be deemed to be the occupier of the factory.

3. Licensing and Registration

As per the Act, the occupier should, at least 15 days before he occupies a place as a factory, will send to the chief inspector of factories a notice containing:

(*i*) Name and address of occupier.
(*ii*) Name and address of factory.
(*iii*) Nature of manufacturing process.
(*iv*) Power to be used.
(*v*) Number of workers likely to be employed.
(*vi*) Any other relevant information.

So before starting the factory the information is to be sent and permission is to be taken.

4. Health

Section 11 to 23 contained in chapter III of the Factories Act 1948 deal with the health of workers in a factory. These provisions are as follows

(*i*) ***Cleanliness:*** Every factory must be kept clean by following methods:

(*a*) Removal and disposal of dirt and refuse from floors, benches etc. everyday.

(*b*) Washing the floors of work room at least once every week using disinfectant.

(*c*) Effective means of drainage to avoid collection of water.

(*d*) Painting is to be done at least once in 5 years.

(*e*) Whitewashing is to be carried out at least once in 14 months.

(ii) Disposal of waste and effluents. Effective and suitable arrangements should be made for the disposal of wastes and effluents due to the manufacturing process.

(iii) Ventilation and temperature. Effective and suitable provision shall be made for maintaining in every work room :

(*a*) adequate ventilation by fresh air circulation.

(*b*) suitable temperature to provide conditions of comfort and prevent injury to the health of workers.

(iv) ***Dust and fumes.*** Employer shall take necessary steps to keep work room free from dust or offensive or injurious to the health of workers.

(v) ***Artificial** Humidification.* In all factories like textile factory, in which the humidity is to be artificially increased, the state government may make the rules for the following:

(*a*) Prescribing standards of humidification

(*b*) Regulating the methods for artificially increasing the humidity of air.

(*c*) Directing prescribed tests for determining the humidity of air.

(*d*) Methods to be adopted for securing adequate ventilation and cooling of the air in the work rooms.

The water used for humidification shall be from a source of drinking water.

(vi) Over-Crowding.

(*a*) No work room will be overcrowded to an extent, which is injurious to the health of workers.

(*b*) The minimum space provided for a worker should be 500 cubic feet (14 m^3).

(*c*) The chief inspector may post a notice in the work room indicating the maximum number of workers who may be employed in that room.

(vii) ***Lighting.*** Every part of the factory must be provided with sufficient and suitable lighting, natural or artificial or both. Effective provision must be made for the prevention of glare and the formation of shadows likely to cause eyestrain. The government may prescribe standards of sufficient and suitable lighting.

(viii) ***Drinking Water.*** Arrangements shall be made to provide drinking water. The points for supply of drinking water should not be situated within 20 feet of any washing place, urinal or latrines. Factories employing over 250 workers are required to provide cool drinking water during hot weather.

(ix) ***Latrines and Urinals***

(*a*) Sufficient latrine and urinal accommodation is to be provided.

(*b*) Separate enclosed accommodation shall be provided for male and female workers.

(*c*) These should be well lighted, ventilated and maintained clean at all times.

(*x*) ***Spittoons***

(*a*) There should be a sufficient number of spittoons at convenient places and maintained in clean hygienic condition.

(*b*) No worker shall spit except in a spittoon.

5. Safety

Section 21 to 41 of the Act contain the provisions relating to the safety of the worker. These are a must and it is the duty of the occupier of every factory to comply with them. Important provisions on safety are given below.

(*i*) *Fencing of Machinery.* The following parts shall be securely fenced unless they are in such a position as to be safe for each worker.

(*a*) Moving parts of prime movers.

(*b*) The head race and tail race of water wheel and water turbine.

The following rotating parts should be fenced by safe guards.

(*a*) Parts of electric generator motor etc.

(*b*) Parts of transmission of machinery.

(*c*) Every dangerous part of any other machinery.

(*ii*) ***Work on or near machinery in motion:***

(*a*) Any part of machinery requiring the examination while it is in motion shall be examined only by a specially trained adult male worker wearing tight fitting clothing.

(*b*) No woman or young worker shall be permitted to clean, lubricate or adjust any part of a moving machinery which involves a risk of injury

(*iii*) *Employment of young persons on dangerous machines.* Noyoung person shall work at any dangerous machines unless,

(*a*) he has been fully instructed as to the dangers arising in connection with the machines and the precautions to be taken.

(*b*) he has received sufficient training in work at the machine.

(*iv*) ***Hoists and Lifts.*** These should be of good construction, sufficient strength and well protected. Hoists and cranes meant for carrying persons shall have at least two ropes of chains separately connected with the cage.

(*v*) ***Lifting machines, chains, ropes and lifting tackles.*** In every factory these should be of good construction, sound material, adequate strength and free from defects. These must be checked by a competent person at least once a year.

(*vi*) ***Pressure plants.*** Measures are to be taken that pressure does not exceed the safe working pressure.

(*vii*) ***Floors, stairs and means of access.*** These should be of sound construction, properly maintained and kept free from obstructions and substances likely to cause persons to slip.

(*viii*) ***Pits, sumps, opening in floor etc.*** In every factory, every fixed vessel, sump, tank, pit or opening in the ground or in a floor, which may be a source of danger, shall be either covered or securely fenced.

*(ix) **Excessive weights.*** No worker will be asked to lift, carry or move any load so heavy as to cause him an injury.

(x) *Protection of eyes.* To protect the eyes of workers from flying objects and particles (like grinding, scale removal etc.) or from exposure to welding rays, each worker shall be provided with effective screens or suitable goggles.

(xi) *Precaution against dangerous fumes.* In any factory, no person shall enter or be permitted to enter any chamber, tank, vat, pit, pipe or other confined space in which dangerous fumes are likely to be present.

*(xii) **Explosives or Inflammable dust, gas etc.*** If a manufacturing process is producing dust, gas, fumes or vapour which can explode on ignition,

(*a*) the plant should be effectively enclosed.

(*b*) such dust, gas fumes etc. should not be allowed to accumulate.

*(xiii) **Precautions in case of fire.*** Every factory shall be provided with

(*a*) fire warning signals.

(*b*) unlocked doors and openings towards outside the work-room.

(c) free passage ways and easily openable windows.

(*d*) Suitable fire extinguishing equipment is to be provided.

6. Welfare

Welfare measures are adopted to increase the productive efficiency of workers. Important welfare measures as per act are given below.

*(i) **Washing facilities.*** In every factory

(*a*) adequate facilities for washing shall be provided and maintained for the use of workers.

(*b*) washing facilities should be separately provided for male and female workers.

(*c*) washing facilities shall be easily accessible and kept clean.

*(ii) **Facilities for sitting.*** Sitting facilities should be provided for workers who are to work in standing positions so that they may take rest if an opportunity occurs in the course of their work without affecting the work.

(iii) *First-aid appliances*

(*a*) Every factory must have first aid boxes or cupboards equipped with prescribed contents necessary for treatment.

(*b*) A factory employing more than 500 workers shall have a properly equipped ambulance room.

*(iv) **Canteens.*** A canteen shall be provided in each factory in which more than 250 workers are ordinarily employed.

*(v) **Shelters, rest rooms and lunch room.*** In every factory where more than 150 workers are employed, adequate and a suitable lunch room shall be provided and maintained for the use of the workers.

*(vi) **Creches.*** Every factory shall provide clean, well lighted and ventilated rooms for the use of children of women workers, if the number of such women workers exceeds 50. The children should be below the age of 6 years.

(vii) ***Welfare Officers***

(*a*) Every factory employing 500 workers or more shall employ welfare officers,

(*b*) The state government may prescribe the duties, qualifications and conditions of service of welfare officers employed.

7. Working Hours

Main provisions in the act dealing with working hours are as follows

(i) ***Daily Hours***

(*a*) No adult worker shall be required to work in a factory for more than 9 hours.

(*b*) No worker shall be required to work for more than 5 hours continuously that is after 5 hours the worker shall have a rest for at least half an hour.

(*c*) No women shall be employed except between the hours of 6 A.M. to 7 P. M.

(*ii*) The hours of work of on adult worker shall be not more than 10½ hours on any day. This includes the hours of work and rest.

(iii) ***Weekly hours.*** An adult worker shall be required to work in the factory for not more than 48 hours a week.

(*iv*) ***Weekly holidays.*** There must be one holiday in every week.

(*v*) ***Extra wages for overtime.*** If a worker works for more than nine hours on any day or more than 48 hours in any week, he will be paid overtime wages which are two time than the daily wages.

(*vi*) ***Register of adult workers.*** A register of adult workers will be maintained and it is to be available for inspection during working hours

8. Employment of young persons

(*i*) A child who has not completed his fourteenth year is prohibited from working in any factory. No exemption can be given to this.

(*ii*) A child who has completed his fourteenth year or an adolescent shall not be required or allowed to work in any factory unless he is given a certificate of fitness by a competent surgeon.

(iii) No child shall be permitted to work

(*a*) for more than four and a half hours on any day, and

(*b*) during the night (*i.e.* 10 P.M. to 6 A.M.)

9. Leave

Every worker who has worked for 240 days or more during a calender year shall be allowed during the subsequent calender year, leave with wages for a number of days at the rate of

(*a*) one day every 15 days of work performed by a child worker.

(*b*) one day for every 20 days of work performed by an adult worker.

Annexure—II

INDIAN BOILER ACT, 1923

The Indian Boilers Act, 1923 consolidates and amends the law relating to Steam Boilers. The Act, however, shall not apply incase of any boiler or steam pipe in any steamship or boiler belonging to Army, Navy or Air Force.

Definitions of Terms

(*i*) "*Boiler*" means any closed vessel exceeding 22.75 litres (5 gallons) in capacity which is used expressly for generating steam under pressure and includes any number of mountings or other fittings attached to such vessel, which is wholly or partly under pressure when steam is shut off. No boiler plate or walls of the vessel shall be less than 6 mm in thickness.

(*ii*) "*Economiser*" means a arrangement used for the purpose of recovery of waste heat.

(*iii*) "*Feed Pipe*" means any pipe or connected fitting wholly or partially under pressure through which feed water passes directly to a boiler and which does not form an integral part of the boiler.

(*iv*) "*Steam Pipe*" means any main pipe exceeding 75 mm in internal diameter through which steam passes directly from a boiler to a steam engine or turbine.

(*v*) "*Accident*" means an explosion or damage of a boiler or steam-pipe.

Registration of Boilers

Every owner of the boiler is required to register his boiler with the Chief Inspector of Boilers. He must submit an application accompanied by the Manufacturer's Drawings of Boiler. On receipt of the application, the Inspector shall fix up a date for the inspection of Boiler and intimate the owner regarding the date of Inspection.

The Inspector shall perform the inspection of the boiler and test it hydraulically at the maximum pressure on which the boiler can be used. The inspector recommend some alterations of fittings and use of insulation on the pipes carrying steam. If the Inspector is satisfied that the boiler is fit for operation, he may recommend to the Chief Inspector of Boilers, for the issuance of certificate. The certificate shall be issued for a maximum period of one year. The maximum pressure at which the boiler can be operated will be recorded on the certificate, issued for operation of the boiler.

Renewal of Certificate

A certificate authorising the use of boiler, shall cease to be in force;

(*i*) On the expiry of the period for which it is granted; or

(*ii*) When any accident occurs to the boiler; or

(*iii*) When the boiler has been transferred from one place to another; or

(*iv*) When any structural alteration, addition or renewal is made in the boiler; or

(*v*) On the communication to the owner of the boiler prohibiting its use on the ground that any of the steam pipe or part is in an unserviceable condition; or

(*vi*) When the Inspector is of opinion that the boiler is not fit for desireduse.

Alterations and Renewals to Boiler

No structural alterations, addition or renewal shall be made to any boiler unless such alterations, additions or renewal has been sanctioned in writing by the Chief Inspector of Boilers.

Alterations and Renewal to Steam Pipes

Before the owner on any boiler makes any structural changes, addition or renewal of any attachment to the boiler, he shall send a report to the Chief Inspector alongwith details for such changes.

Restrictions

1. A boiler cannot be used unless it is registered and a certificate is obtained to that effect from the boiler Inspector.
2. If it is transferred from one state to another, the owner is not allowed to use it unless it is reported to the registering authority in the concerned state.
3. It cannot be used at pressures higher than prescribed by the competent authority issuing the certificate.
4. Only a person having a competent certificate is allowed to operate the boiler.

Duties of Boiler Inspector

1. The boiler inspector should ensure that boilers are registered.
2. He should see that boilers are working according to the provisions of the boiler Act.
3. He should check and examine boilers, their parts, mountings and accessories at regular intervals of time.
4. He should advise the owner of boilers in the matters of boiler mounting, cleaning and their proper maintenance.

Duties of Chief Boiler Inspector

1. He has to maintain record of all registered boilers under his jurisdiction.
2. He has to examine the reports submitted by Boiler Inspectors working under him.

3. He has to decide whether certificate for operation of boiler is to be issued or not.

Boiler Mountings

Every boiler shall be provided at least with the following in order to ensure efficient and safe operation:

(*a*) Two safety valves.

(*b*) Two pressure gauges to indicate the pressure in the boiler.

(*c*) Water Level Indicator.

(*d*) A steam pressure gauge.

(*e*) A steam stip valve.

(*f*) A feed check valve.

(*g*) A blow off cock.

(*h*) A fusible plug for safety of boiler.

(*i*) A manhole or such mud holes and sight holes as are necessary for effectively cleaning the boiler.

(*j*) An additional valve at the end of superheater.

❑❑❑

Annexure—III

ELECTRICITY SAFETY ACT, 1910

(*a*) **General Terms:**

(*i*) **Area of supply:** It means the area within which alone a licensee is for the time being authorised by his license to supply electrical energy.

(*ii*) **Consumer:** It means any person who is supplied with energy by a license or the Government or by any other person engaged in the business of supplying energy to the public under this Act.

(*iii*) **Distributing Main:** It means the portion of any main with which a service line is or is intended to be immediately connected.

(*iv*) **Electric supply line:** It means a wire, conductor or other means used for transmitting or distributing energy.

(*v*) **Energy:** It means electric energy generated, transmitted, supplied or used for any purposes except the transmission of a message.

(*vi*) **"Licensee":** It means any person licensed to supply energy.

(*vii*) **Main:** It means any electric supply line through which energy is or is intended to be supplied to the public.

(*viii*) **Overhead Line:** It means an electric supply line which is placed above ground and is in the open air with the help of poles.

(*ix*) **'Public Lamp':** It means an electric lamp used for the lighting of any street/lane/bazar etc.

(*x*) **"Service line":** It means any electric supply line through which energy is intended to be supplied:

(*i*) to a single consumer either from a distributing main, or

(*ii*) from a distributing main to a group of consumers on the same premises or on adjoining premises supplied form the same point of distributing main.

(*xi*) **Street:** It includes any way, road, lane, square, passage or open space. Whether a thoroughfare or not, over which the public have a right of way and also the roadway and footway over any public bridge of causeway.

(*xii*) **Works:** It includes electric supply line and any building, plant, machinery, apparatus and any other thing required to supply energy to the public.

2. Grant of Licenses

The State Government may on application made in the prescribed form and on payment of prescribed fee grant a licence to any person to supply energy in any specified area and also to lay down or place electric supply lines for the transmission of energy.

The grant of licence to any licensee for any purpose shall not in any way hinder or restrict the grant of a licence to another person within the same area of supply for a similar purpose.

3. Revocation of Licenses

The State Government may if in its opinion the public interest so requires revoke a licence in any of the following cases, namely—

(*i*) Where the licensee, in the opinion of the State Government, attempts wilful and unreasonably prolonged default in performing any function required of him under this Act;

(*ii*) Where the licensee goes against any of the terms and conditions under this Act;

(*iii*) Where the licensee fails to prove to the satisfaction of the State Government that he is in a position fully and efficiently to discharge the duties and obligations required from him by his licence or he fails to furnish the security required by him;

(*iv*) Where in the opinion of State Government the financial position of the licensee is such that he is unable to discharge his duties efficiently;

(*v*) Where a licensee, in the opinion of the State Government, has made default in complying with any direction under this Act.

State Government shall not revoke any lincese without giving a three months' notice in writing stating the grounds on which it is proposed to revoke the license and has considered any cause shown by the licensee within the period of that notice, against the proposed revocation.

4. Amendment of Licenses

State Government on the application of licensee or otherwise may make amendments in the terms and conditions of the license, where in its opinion public interest permits.

5. Provisions where License is Revoked

The Government shall serve a notice of revocation upon the licensee and shall fix a date on which the revocation shall take effect. Where the State Government revokes the license of a licensee. The State Government shall either purchase the undertaking itself or arrange to sell to the State Electricity Board or to some other local authority.

6. Determination of Purchase Price of an Undertaking

Where an undertaking of a licensee is sold, the purchase price of the undertaking shall be the market value of the same prevailing at the time of its purchase. The market value of the underaking shall include all lands, building and plant of the licensee required for the undertaking.

7. Annual Accounts

Every licensee shall prepare annual statement of accounts of its undertaking and submit a copy of the same to the State Government for their inspection. Further, the licensee shall keep copies of such annual statement at his office and sell the same to any applicant at a price not exceeding five rupees per copy.

8. Laying of Lines in Streets, Railways, etc.

The licensee shall have the power to lay down or place electric supply lines subject to the condition that he must take the consent for such work from the local authority (municipal committees or corporations) or the owner or occupier of property in the area:

(*a*) to open and break up the soil and pavement of any street, railway or tramway;

(*b*) open and break up any sewer, drain or tunnel under any street, railway or tramway; with prior information with concerned authority;

(*c*) lay down and place electric supply lines;

(*d*) repair, alter or remove the same;

(*e*) do all other acts essential for the due supply of energy.

9. Alteration of Pipes or Wires

The licensee is authorised to change the position of any pipe or wire under or over any place which he is authorised to open or break up, if such pipe of wire is likely to interfere with the exercise of his powers under this Act. However, he has to give a notice of at least one month alongwith details of alterations to be carried out, to the owner for his information and any objection is to be given.

10. Filling up of Drains and Sewers in the Street

Where any person opens or breaks up the soil of pavement of any street, railway or tramway of any sewer, drain or tunnel, he shall immediately cause the part opened or broken up to be fenced and guarded properly. He shall also fix lights near the part opened or broken for the warning of passengers before the sunset so that any person may not meet accident.

Further, he shall fill in the ground and make good the soil or pavement, or the sewer, drain or tunnel immediately after the work is completed and shall carry away the garbage from the site.

11. Power of Licensee to Enter Premises

The licensee or any person duly authorised by him has the power to enter the premises of any person for the following lawful objectives:

(*a*) For inspecting, testing, repairing or altering the electric supply lines, meters, fittings regarding the supply of energy.

(*b*) For ascertaining the amount of energy supplied or electrical quantity contained in the supply.

(*c*) For removing the meters or other fittings where the electric energy is no more needed.

12. Charges for Energy

The licensee may charge for energy supplied by him to any consumer on the following basis.

(*i*) By the actual amount of energy so supplied

(*ii*) By the electrical quantity contained in the supply

(*iii*) By such methods as approved by State Government.

13. Discontinuance of Supply

Where any person fails to make payment of charges for energy due to him in connection with the supply of energy, the licensee may cut off the supply of electricity after giving him due notice of seven days. The supply will remain discontinued until the amount together with expenses incurred by him in cutting off and reconnecting the supply are paid to the licensee.

14. Control of Transmission and Use of Energy

No body is authorised to use or transmit energy at a rate exceeding 250 watts and 220 volts in any street or place except by the licensee or duly authorised person.

15. Protection of Railways, Aerodromes etc.

No body shall injure the Railway, tramway, aerodrome or water work installation while supplying or using the energy. He will also not interfere or obstruct the traffic on Railway, Tramway or at Aerodrome.

16. Protection of Telegraphic and Telephonic Lines

Every person while generating, transmitting, supplying or using energy shall take precautions in constructing, laying down and placing his electric supply lines so as not to interfere with the working of any line used for the purpose of telegraphic, telephonic or signalling communication.

Central Electricity Board

A board called Central Electricity Board shall be constituted to exercise the following powers:

(*i*) To make rules to regulate the generation, transmission and utilization of energy

(*ii*) To provide for preparation and submission of accounts by the licenses in the requisite performs

(*iii*) To provide for securing a regular, constant and sufficient supply of energy by the licensee to the consumers

(*iv*) To provide for as required by them protection of telegraph lines and magnetic observation laboratories from being injuriously affected due to generation, transmission, supply for use of energy

(*v*) To authorize inspectors to enter and inspect any place and carry out test

(*vi*) to regulate publication of notices regarding energy transmission and distribution

(*vii*) To prescribe qualifications to be required for electrical inspectors.

Protective Equipment

The following protective equipment should be kept ready near the electrical installation for safety purpose:

(*i*) Fire buckets containing dry sand for immediate use as a means of extinguishing fire

(*ii*) Fire extinguishers at convenient places should also be provided at generating stations and substations meant for distribution

(*iii*) First aid boxes must also be provided alongwith a person trained in first aid treatment

(*iv*) Works of electrical installation and repairs should only be carried out by *Licensed Electrical Contractors.*

(*v*) Every installation being used in supply system should be periodically checked and tested.

Criminal Offences

1. **Theft of Energy:** Whosoever dishonestly by obstructs, consumers, or users energy shall be deemed to have committed theft within the meaning of Indian Penal Code.
2. Whosoever intentionally causes energy to be diverted, cuts or injures any supply line shall be punishable with imprisonment upto 2 years or with a fine or Rs. 1000 or both.
3. If a licensee supplies energy outside the area of supply or if he without reasonable causes discontinues the supply of energy he shall be punishable with a fine upto Rs. 1000.
4. Penalty for interference with meters.

   Whosoever intentionally injures any meter or wilfully alters the index of any meter or prevent such a meter from duly registering the consumption of energy is defaulter and shall be punishable with fine upto Rs. 500.
5. Whosoever intentionally extinguishes any public lamp shall be punishable with imprisonment for a term which may extend to 6 months or with fine upto Rs. 100 or with both.
6. Whosoever negligently causes energy to be wasted or diverted or negligently breaks or throws down or damages any electric supply, post, pole or lamp etc. connected with supply of energy shall be punishable with fine upto Rs. 200.

❑❑❑

Annexure—IV

THE WATER (PREVENTION AND CONTROL OF POLLUTION) ACT, 1974

The salient features of this act are given below:

Object : This Act has been passed on water pollution.

(*i*) The prevention and control of water pollution.

(*ii*) The maintaining or restoring wholsomeness of water.

(*iii*) The establisying of Boards for prevention and control of water pollution.

(*iv*) Fine is generally little as compared to expenditure involved in compliance of standards.

(*v*) Information about the polluting establishments is not readily available to other citizens.

Application : The Act is not applicable in whole of India. It applies to those states which have requested the parliament to pass it and those which may subsequently adopt it. However, it is applicable in all the Union Territories.

Definition of Water Pollution: Since the Act is to prevent and control pollution of water, therefore, it has given elaborate definition of pollution, which are given below :

Pollution means:

1. Contamination of water
2. Alteration of the physical, chemical, or biological properties of water.
3. Discharge of any sewage or trade effluents or any other liquid, gaseous or solid substance into the water (directly or indirectly), which is likely to
   (*a*) Create a nuisance, or
   (*b*) Render it harmful or injurious :
      (*i*) To public health or safety or to domestic, commercial, industrial, agricultural or other user; or
      (*ii*) To the life and health of animals or plants or of acquatic organisms.

Central Board. The Act provides for constitution of central board for prevention and control of water pollution. The central board shall be directly under the control of Government of India and shall work independently.

The Central Board shall consist of the following members.

(*a*) One full time Chairman.

(*b*) Not more than five officials nominated by Central Government.
(*c*) Not more than five persons nominated by the Central Government from amongest the members of State Boards.
(*d*) Not more than three non-officials nominated by the Government to represent interests of agriculture, fishery, industry, and trade etc.
(*e*) Two persons nominated by Government to represent companies or corporations of the Government.
(*f*) One full time member secretary.

State Boards. The State Government may constitute a State Pollution Control Board. The State Boards are directly under the control of central board or the State Government varying from state to state. These boards are responsible for dealing with local problems as per the direction of central boards or State Government.

A State Board shall consist of the following members:
(*a*) One full time Chairman.
(*b*) Not more than five officials nominated by the State Government.
(*c*) Not more than five persons nominated by the State Government from amongst the members of the local authorities functioning with in the state.
(*d*) Not more than three non-officials nominated by the Government to represent interests of agriculture, fishery, industry and trade, etc.
(*e*) Two persons nominated by Government to represent Companies or Corporations of the Government.
(*f*) One full time Member Secretary.

Terms and Service Conditions of Members of Boards: The terms and conditions of service of Member-Secretary shall be as may be prescribed by the Government. Other members shall hold office for a term of three years. Official members shall cease to be members when they cease to be such officials. A member shall be eligible for renominations. Central/ State Government may remove a member of the Central/State Board at any time by giving him reasonable opportunity.

The terms and service conditions of the Chairman shall be as may be prescribed by the Government. The Chairman may resign by addressing his resignation to the Government. A member may resign by addressing resignation to the Chairman.

Seat of a member shall fall vacant if he fails to attend three consecutive meetings of the Board without sufficient reasons or if he incurs any disqualification.

A person nominated to fill a casual vacancy shall hold office for the remaining term.

At least one meeting of a Board must be held in every three months.

Functions of Central Board : Functions of Central Board are :
(*i*) Cleanliness of streams and wells in different areas of the states.
(*ii*) To advise the Central Government regarding water pollution control.

(*iii*) To Co-ordinate activities of State Boards and resolve disputes among them.

(*iv*) To plan and organise training of persons in the field of water pollution.

(*v*) To utilise the services of mass media in educating public about causes, effects, prevention and control of water pollution.

(*vi*) To lay down standards for quality of water.

(*vii*) To establish or recognise laboratories for analysis of water samples etc.

(*viii*) To prepare manual, codes or guides for treatment of effluents.

(*ix*) To carry out and sponsor investigations and research related to water pollution.

(*x*) To look after pollution control in union territories.

Functions of State Board: The Functions of State Board are :

(*a*) To advise State Government regarding water pollution control or location of industries.

(*b*) To plan programme for prevention and control or abatement of water pollution.

(*c*) To conduct and encourage research relating to water pollution.

(*d*) To collect and disseminate information related to causes, prevention and control of water pollution.

(*e*) To seek guidance and training of persons connected with prevention, control or abatement of water pollution from Central Board.

(*f*) To inspect effluents and treatment plants.

(*g*) To lay down standards of effluents.

(*h*) To evolve economical and reliable methods and lay down standards for treatment of sewage and trade effluents.

(*i*) To evolve methods for utilisation or disposal of treated sewage and effluents.

(*j*) To perform such functions as may be entrusted by Central Board or State Government.

(*k*) To establish or recognise laboratories for analysis of samples.

Power of Board

1. Every industrial house must obtain consent from State Board about the method of treatment and disposal of sewage or trade effluent, amount, amount contents and place or discharge into a stream or well. A new consent is required for any addition or alteration.

2. It has the power to obtain information from individuals about removal or discharge of water, construction installation or operation of any disposal system in establishment where industry or trade is carried out.

3. To take samples of any sewage or trade effluent, water from any stream and well.

4. To enter any premises and inspect any plant, record, register, document and condut search of a place if the authority has reason to believe that an

offence under this act has been committed in residential areas and the entry can be only at reasonable hours.

5. In case of any emergency or pollution causing accident, the Board can take remedial action in writing that may include restraining or prohibiting discharge of pollution.

6. Normally the Board makes an application to court after detecting an offence for giving injunctive relief, restraining an establishment from causing pollution of water in stream or well.

7. Under the amended Act the State Board can issue directions to any person, officer or authority about regulation prohibition or closure of process, operation or industry. It can stop or regulate supply of water, electricity or any other service.

Penalties

Punishment, in case of failure to provide information or comply with directions of the Board (section 20), a person on conviction by Court, is liable to be punished with imprisonment upto three months or/and a fine of upto 10,000 rupees. In case the failure continues he can be punished with an additional fine of upto Rs. 5000 per day during which such failure continues. A person convicted of causing pollution due to nonobservance of standards is liable to be inprisoned for 1 *Vz* years to six years or and a fine of upto ten thousand rupees. Continued failure invites an additional fine of 5000 rupees per day.

Limitation

(*a*) The Act does not provide for involvement of people or voluntary agencies in its implementation.

(*b*) Power of the Board is largely advisory in nature. Even the consent application would be deemed to have been cleared if the authorities do not communiate the same within four months.

(*c*) The Board can take remedial measures only in the event of accidents or unforseen occurrences.

(*d*) Normally the Board is to take recourse to court of law which takes a long time to come to the conclusion. By that time sufficient harm is done to water courses.

(*e*) Provision for giving 60 days notice by a private agency excludes the chance for immediate remedial action.

❑❑❑

Annexure—V

THE AIR (PREVENTION AND CONTROL OF POLLUTION) ACT, 1981

Object

The salient features of this Act are given below:

The main object of this Act is to prevent, control and abate air pollution.

Application

Unlike the water pollution control Act, this Act is applicable in whole of India. This Act came into force from may, 1981.

Definitions

(*i*) Air pollution means the presence in the atmosphere of any air pollutant.

(*ii*) Air pollution means any solid, liquid or gaseous substance (including noise) present in the atmosphere in such quantity as may be injurious to human beings, other living creatures or plants or property or environment.

Boards. In order to have integrated approach for tackling the pollution problems, the Act provides that Boards established under the Water Pollution Control Act, 1974 will also act as Air Pollution Control Boards under this Act. But in those states in which State Board for Water Pollution Control are not established separate State Board for Air Pollution Control will be established.

Functions of Central Board

The Functions of Central Board are:

1. To improve quality of air in the country.
2. To prevent or control air pollution in the country.
3. To advise the Central Government and State Board on any matter related to the improvement of air quality or control of air pollution.
4. To plan and execute programmes for control of air pollution on nation wide scale.
5. To arrange training programmes for prevention and control of air pollution.
6. To co-ordinate activities of the State Boards.
7. To collect and publish technical and statistical data on air pollution.
8. To lay down standards for quality of air.
9. To provide on arrange sponsorship for investigations and research relating control and prevention of air pollution.
10. To establish laboratories.

Functions of State Board

Functions of State Board are given below:

1. To plan programmes for prevention and control of air pollution.
2. To co-operate with Central Board to collect and distribute information regarding air pollution control Programmes.
3. To advise the State Government regarding matters related to air pollution within the state.
4. To inspect various industrial plants and control equipment and give necessary orders to prevent and control air pollution.
5. To inspect air quality in air pollution control areas from time to time.
6. To Suggest necessary steps to be taken to control air pollution in air pollution control area.
7. To lay down standards for emission of air pollutants within the state (in consultation with Central Board.)
8. To Set up laboratories.

Power of the Board

1. To recommend to State Government to declare any area or areas within the state as air pollution control area or areas. The whole of Punjab has been declared as air pollution control area.
2. On the recommendation of State Board, the State Government can direct any industry to use only approved fuel and approved appliances. It can prohibit the burning of any material.
3. Every industrial establishment has to move application of consent. After making inquiry and studying details, the State Board shall refuse or grant the consent subject to such conditions like specification of control equipment, its periodic checking, alterations a"nd replacements, and specifications of chimney.
4. The board can restrict industrial operations in air pollution control areas.
5. It can withdraw consent given to any industry and impose new conditions.
6. The Board can give instruction to authorities under Motor Vehicles Act regarding standards for automobile emission.
7. Board officials have the power to enter any premises at all reasonable times and inspect compliance with requirements of the Act.
8. To call information from any industry with regard to pollutants emitted into atmosphere.
9. To take air or emission samples for analysis.
10. In case of any emergency due to accidents or unforseen events causing pollution in pollution control area, the Board can take remedial action in writing.

Assessment of Air and Water Pollution from Industries

Every industrial enterprise has to file all data pertaining to the assessment of air and water pollution as per proforma given below.

ANNEXURE

Questionnaire for Assessment of Air and Water Pollution from Industries:

(**Source:** Central Board of Prevention and control of Water Pollution, North-Western Regional Office, Chandigarh)

DATA SHEET

To be filled in by the Central Board

Area Code : ....

To be filled in by the industry

A. BASIC INFORMATION

1. Name of the Industry ....
   (*a*) Registered Address: ....
   (*b*) Factory Address ....
2. Person Contacted
   (*a*) Name: ....
   (*b*) Designation: ....
   (*c*) Office/Factory Address: ....
   (*d*) Telephone No.: ....
   (if any)
3. Indicate existence in surrounding of industry/factory's

Residential building	Industries
Commercial building	Others (specify)

4. (*a*) Year of Commissioning of Industry. ....
   (*b*) Total area of the factory premises ....
   (*i*) Built up area ....
   (*ii*) Area available for treatment plants ....
   (*c*) Number of personnel working in factory/industry ....

	Shift-I	Shift-II	Shift-III	Gen-Shift
• Working Hours	....	....	....	....
• No. of Hours	....	....	....	....
• No. of Hours in each shift	....	....	....	....
• Total No. of workers	....	....	....	....

Close day

(*d*) Are any personnel residing in the premises (Give number): Yes/No

(*e*) Operating period: Seasonal (Specify period/whole year.

5. Product Manufactured

	Name of product	Qty. per month	Installed capacity
(*a*)			
(*b*)			
(*c*)			
(*d*)			
(*e*)			
(*f*)			

6. Raw materials (other than fuel and chemicals)

	Name	Source	Qty. per month
(*a*)			
(*b*)			
(*c*)			
(*d*)			
(*e*)			
(*f*)			

7. Fuel consumption

	Name	Source	Qty. per month
Coal (specify grade)			
Oil			
Natural gas			
Wood			
Others (specify)			

8. Manufacturing Processes Involved

B. Emission Inventory of Air Pollutants

1. Melting Furnaces:
   Pot/Revoberatory/Electric Induction or resistance/Crucible/Cupola/Electric Arc/Other Metallic (Describe)
   (*a*) No. of such melting devices:
   (*i*) Installed .... (*ii*) In use ....
   (*b*) Date of installation ....
   (*c*) Capacity in tonnes/hour ....
2. Boilers:
   (*a*) No. of boilers ....
   (*i*) Installed .... (*ii*) In use ....
   (*b*) Capacity in tonnes/hour ....
   (*c*) Date of installation ....
   (*d*) Type/make ....
   (*e*) Capacity ....
   (*f*) Method of fuel charging
   (*i*) Oil/gas burners
   (*ii*) Grate Charges (manual/mechanical)
   (*iii*) Pulverized Coal Chargings
   (*g*) Types of burner Axial/Tangential
   (*h*) (*i*) Type:
   (*ii*) Quantity:
3. Stack Data:
   (*a*) No. of Stack ....
   (*b*) Material of Construction:
   (*c*) Height (*i*) above roof: .... (*ii*) above ground: ....
   (*d*) Diameter: (*i*) Bottom .... (*ii*) Top ....
   (*e*) Draughts : (*i*) Bottom .... (*ii*) Induced ....
   If induced, capacity of fan.

4. Any other process emitting chemical vapour of fuel gas:
5. Emission control equipment for air pollutants. Tick the condition prevailing.
   Existing/not existing/proposed/not proposed
   If existing or proposed:
   (*a*) Nature of pollution control equipment ....
   (*b*) Designed by ....
   (*c*) Installed by ....
   (*d*) Capacity ....
   (*e*) Efficiency ....

C. ASSESSMENT OF WASTE WATER DISCHARGE

1. Source of water supply and quantity used:

	Public Health Supply	Own Qty. (m^3/day)	
(*a*) Processes	....	....	....
(*b*) Cooling	....	....	....
(*c*) Equipment and floor washing	....	....	....
(*d*) Residential* Colony, if any	....	....	....
(*e*) Canteen/toilets/ gardening	....	....	....

2. If tubewell is used, please indicate
   (*a*) Depth ....
   (*b*) Diameter ....
   (*c*) Pump H.P. ....
   (*d*) Discharge (m^3/hrs) ....
   (*e*) Pumping hours/day ....
3. Provision of water storage tank: Yes/No
4. If yes: specify
   (*a*) Treatment, if any given to the water before use and quantity treated, m^3/day
   (*b*) Chemical and their quantities used for conditioning or corrosion control in cooling system.
5. Quantity of wastewater generated and treated:

Types of Wastewater	**Generated (m^3/day)**	**Treated (m^3/day)**
(*a*) Process		
(*b*) Cooling		
(*c*) Equipment and floor washings		
(*d*) Canteen and toilets		
(*e*) Residential colony		

6. Brief description of the treatment plant:

(*a*) Types of plant
(*b*) Area occupied by the plant
(*c*) Contracting agency/constructed by
(*d*) Consulting agency
(*e*) Year of construction ....
Data of commissioning ....
Subsequent expansion if any ....

(*f*) Unit Processes	Existing	Size Design Parameters
— Equalisation tank	Yes/No	
— Chemical Treatment (neutralisation and/or heavy metal removal)	Yes/No	
— Primary Settling	Yes/No	
— Activated Sludge (Type)	Yes/No	
— Oxidation ditch	Yes/No	
— Aerated lagoon	Yes/No	
— Secondary Settling	Yes/No	
— Sludge Digestion	Yes/No	
— Sludge drying body/ Vacuum filter	Yes/No	

(Signature of the person Contacted)
Date:

(Signature of the Investigating officer)
Date:

❑❑❑

Annexure—VI

THE INDIAN PARTNERSHIP ACT, 1932

1. Partnership deed

Partnership is an agreement between persons to carry on a business. The agreemente entered into between partners may be either oral or writte. But, it is always desirable to have a written agreement so as to avoid misumderstanding and unnecessary litigations in future. When the agreement is in written form, it is called 'Partnership Deed.' It must be duly signed by the partners, stamped and registered. Any alteration in partnership deed can be made with the mutual consent of all the partners.

Although it is left to thedoice of the partners of the firm to decide themselves as to what should be mentioned in their partnership deed, yet a partnership deed generally contains the following:

1. Name of the firm.
2. Nature of the business.
3. Names of partners.
4. Place of the business.
5. Amount of capital to be contributed by each partner.
6. Profit sharing ratio between the partners.
7. Loans and advances from the partners and the rate of interrest thereon.
8. Drawings allowed to the partners and the rate of interest theoreon.
9. Amount of salary and commission, if any, payable to the partners.
10. Duties, powers and obligations of partners.
11. Maintenance of acounts and arrangement for their audit.
12. Mode of valuation of goodwill in the event of admission, retirement and death of a partner.
13. Settlement of accounts in the case of dissolution of the firm.
14. Arbitration in case of disputes among the partners.
15. Arrangements in case a partner becomes insolvent.

2. Registration of the Firm

Under the Indian partnership Act, 1932, the registration of the firm is not compulsory. An unregistered firm suffers from certain limitations, hence the registration of the firm is desirable. Registration can be done at any time.

The procedure for registration is as follows:

The firm will have to apply to the registrar of firms of the respective State Government in a prescribed application form. The form should be duly signed by all the partners. The application form should contain the following information:

1. The firm name.
2. The name of business place.
3. Names of other places, if any, where the firm is carrying on its business.
4. Date of commen cement of business.
5. Date when each partner joined the firm.
6. Full names and permanent addresses of all the partners.
7. The duration of the firm, if any.

When the Registrar of firms is satisfied that all formalities relating to registration have been fully complied with, he makes an entry in the Register of Firms. Thus, the firm is considered to be registered. The register of Firm remains open for inspection of payment of prescribed fee for the purpose.

3. Dissolution of Firm

There is a difference between the dissolution of partnership and dissolution of firm. Dissolution of partnership occurs when a partnership occurs when a partner ceases to be associated with the business, whereas dissolution of firm is the winding up the business. In other words, in case of dissolution of partnership, the business of the firm does not come to an end but there is a new agreement between the remaining partners. But in case of dissolution of firm, the business of the firm is closed up. In brief dissolution of partnership does not imply the dissolution of firm. ut, dissolution of firm implies dissolution of partnership also.

Following are the various ways in which a firm may be dissolved:

(*i*) **Dissolution by Agreement:** The partnership firm may be dissolved in accordance with a contract already made between the partners.

(*ii*) **Compulsory Dissolution:** A firm stands compulsorily dissolved under the following circumstances:

(*a*) By the adjudication of all the partners or of all the partners but one as insolvent.

or

(*b*) By the happening of any such event that makes the business unlawful.

(*iii*) **Dissolution Due to contingencies:** A firm stands dissolved on the happening of the any of the following contingencies:

(*a*) On expiry of partnership period, if constituted for a fixed period.

(*b*) On completion of the firm's venture for which the firm was formed.

(*c*) On the death of a partner.

(*d*) On the adjudication of a partner as an insolvent.

(*iv*) **Dissolution by Court:** Under any of the following cases, a court may order the dissolution of a firm:

(*a*) Any partner has become of unsound mind.

(*b*) Any partner has become permanently incapable of performing his duties as a partner.

(*c*) A partner's misconduct is likely to affect prejudicially the business of the firm.

(*d*) A partner wilfully commits breach of the partnership agreement.

(*e*) A partner transfers his interest in the firm, but unauthorisedly, to a third party.

(*f*) The business of the firm can be carried on at loss only.

(*g*) It is just and equitable, on the basis of ay other reasonable ground, that the firm should be dissolved.

4. Settlement of Accounts on Dissolution

Settlement of accounts means closure of all accounts in the books of the firm as the firm's business no longer exists. According to Section 48 of the Indian Partnership Act, 1932, the procedure for the settlement of accounts after the dissolution of the firm is as follows:

The assets of the firm are disposed of and the amounts so realised are applied in the following manner:

(*i*) Payment of debts due to the third parties.

(*ii*) Rateable payment of loans and advances made by the partners to the frm.

(*iii*) Payment of partner's capital.

(*iv*) Payment of surplus, if ay, t the partners in their profit sharing ratio.

The losses of the firm on dissolution have to be made up:

(*i*) First out of accumulated past profits.

(*ii*) Then, out of capitals of partners.

(*iii*) Thereafter, out of contributions from the private estates of the partners in their profit sharing ratios.

It is important to mention that the private property of the partner is to be used first to pay his private debts and only the surplus, if any, can be used to pay firm's liabilities. Similarly, firm's assets are first used to pay firm's liabilities. Only surplus can be used to pay the partner's private liabilities.

Annexure—VII

WORKMAN'S COMPENSATION ACT, 1923

The Act seeks to compensate the workers injured. This Act came into existence in 1923. The Act was intended subsequently in 1933, 1938, 1939, 1946, 1958, 1962 and 1976.

This Act is applied to all workers in the factories, mines, docks and railways. It is also applicable to transport workers, telegraph and telephone linesmen and workers in building profession.

Necessity. This Act was necessary due to the possibility of more accidents in the evergrowing complexity industry. The Act provides for cheaper and quicker disposal of disputes relating to compensation. The orker (or his dependents) can claim compensation if the injury has been caused by an accident in the purse of employment. The workers are protected against the financial hardships arising out of accidents.

Important Definitions in the Act:

1. **Dependent:** It means
   (*a*) A widow, a minor son unmarried daughter or a widowed mother.
   (*b*) If wholly or in part dependent on the earnings of the workman at the time of his death. This will include a minor brother and unmarried sister and a widowed daughter-in-law.
2. **Employer:** The term includes the following:
   (*a*) Anybody or persons whether incorporated or not.
   (*b*) Any managing agent of an employer.
   (*c*) Legal representative of a deceased employer.
3. **Disablement:** Disablement is a condition resulting from employment injury of the worker which makes him incapable to work temporarily or permanently.
4. **Temporary Partial Disablement:** It means such disablement as temporarily reducing the earning capacity of a worker which he was capable of, at the time of accident, e.g., fracture of a arm or leg.
5. **Permanent Partial Disablement:** It means such a disablement which permanently reduces the earning capacity of a worker in every employment which he was capable of undertaking at the time of accident, e.g., loss of thumb, amputation of one hand or one leg.

6. **Total Disablement:** It means such disablement which is of permanent nature and percentage loss of earning capacity is 100% in this case.
7. **Wages:** It includes all type of wages, privileges and benefits, bonus is also included in wages for the purpose of calculation of compensation. Dearness allowance and amenities of free water, electricity and quarters, are the privileges included in the wages.
8. **Workman:** It means any person who is employed on monthly wages not exceeding Rs. 1000/-.
9. **List of injuries and Percentage loss of Earning Capacities.** There are 54 injuries mentioned in the heddle I of the Act. Some of which are as follows:

S.No.	Description of Injury	Percentage loss of Earning capacity
1.	Loss of both hands or amputation at higher rates.	100
2.	Loss of a hand and a foot.	100
3.	Double amputation through leg or thigh or amputation through leg or thigh one of the side and loss of other foot.	100
4.	Loss of sight to such an extent as to render the claimant unable to perform any work for which eyesight is essential.	100
5.	Very severe facial disfigurement.	100
6.	Absolute deafness.	100
7.	Amputation through shoulder joint.	90
8.	Loss of a hand or of the thumb and four fingers of one hand or amputation from 4½″ (114 mm nearly) below tip of olecranon.	60
9.	Loss of thumb	30
10.	Loss of thumb and its metacarpal bond.	40
11.	Loss of four fingers on one hand.	50
12.	Loss of three fingers on one hand.	30
13.	Loss of two fingers on one hand.	20
14.	Amputation of both feet.	90
15.	Amputation through both feet proximal to the metatarso-phalangeal joint.	80
16.	Loss of toes of both feet through the metatarasophalan-geal joint.	40
17.	Amputation at hip.	90
18.	Amputation below middle high thigh to 3½″ (89 mm nearly) below knee.	60
19.	Loss of one eye, without complications, the other being normal.	40

20.	Loss of vision of one eye, without complications of disfigurement of eye-ball, the other being normal	30
21.	Index finger whole.	14
22.	Guillotine amputation of tip without loss of bone, middle finger.	5
23.	Middle finger, whole.	12
24.	Guillotine amputation of tip without loss of bone, index finger.	4
25.	Little finger whole.	7
26.	Guillotine amputation of tip without loss of bone, little finger.	2
27.	Great toe, through metatarasophalangeal joint.	14
28.	Great toe part, with some loss of bone.	3
29.	Low part of toe with loss of bone.	1

Salient Features of the Act

1. Application of Act
2. Exceptions
3. Amount of Compensation

1. Application of Act

This Act is applicable to any worker.

(*a*) Who is employed in a factory, employing 10 or more workers.

(*b*) The monthly wages does not exceed Rs. 1000/-.

(*c*) The injury has been caused by an accident and during the course of employment.

(*d*) The accident has resulted in some type of disablement. The employer will have to pay the compensation depending upon the disablement.

2. Exceptions

This Act is not applicable where

(*a*) The injury disables a worker for less than three days.

(*b*) The injury is caused by an accident which occurred while the workman was under the influence of drink or intoxicants.

(*c*) The injury is caused due to wilful disobedience of the rules of the workman.

(*d*) The injury is caused owing to the wilful removal of any safety guard by the workman.

In such cases the employer is not liable to pay any compensation.

3. Amounts of Compensation

The following factors are considered in deciding the amount of compensation:

(*a*) average monthly wages of a workman.

(*b*) The extent of injury.

(*c*) The nature of injury. Resulting in death, permanent, total or partial disablement, and temporary disablement.

(*d*) The amount of compensation payable in case of different injury can be found from the tables.

(*e*) Payment of compensation due to death of a workman or under legal disability shall be deposited by the employer with the commissioner.

(*f*) The injured workman should first of all give in writing a notice of accident to the employer. The notice contains particulars of workman, date of accident, and cause of accident. The injured workman who has submitted a notice shall present himself for the medical examination, if the employer wants. Later on compensation is claimed for payment through the commissioner.

Illustration

1. A workman whose monthly wages exceed Rs. 400/- but not Rs. 500/- sufferers injury by loss of his right hand. The compensation payable to him shall be calculated as follows: (See Table - Rates of Compensation)

Percentage loss of earning capacity by loss of right hand = 60

Amount of compensation in case of permanent total disablement = 29,400

∴ Amount of compensation payable to him

$$= \frac{29,400 \times 60}{100} = 17,640$$

2. If in illustration 1. The workman had lost both his hands. The compensation shall be calculated as follows:

Percentage loss of earning capacity by loss of both hands = 100

From the tables the compensation is = 29,400

Minimum rate of compensation for permanent total disablement and death have been fixed at Rs. 90,000 and 80,000 respectively. Maximum amount for death and permanent total disablement can go upto Rs. 4.56 lakhs and 5.48 lakhs respectively depending on age and wages of workmen.

Rates of Compensation

Monthly wages of the workman injured		Amount of Compensation for		Half monthly payment as compensation for temporary displacement
More than	But not More than	Death	Permanent total disablement	
(1)	(2)	(3)	(4)	(5)
Rs.	Rs.	Rs.	Rs.	Rs.
0	30	3,600	5,400	20.00
30	60	7,200	10,080	20.00
60	90	9,720	14,608	36.00
90	120	11,520	16,128	42.00
120	150	13,500	18,900	48.75
150	200	16,800	23,520	60.00
200	300	18,000	25,200	82.50
300	400	19,200	26,880	100.00
400	500	21,000	29,400	118.75
500	600	21,600	30,240	135.00
600	700	23,100	32,340	148.72
700	800	24,000	33,600	160.00
800	900	27,000	37,800	168.75
900	1000	30,000	42,000	175.00

Annexure—VIII

PAYMENT OF WAGES ACT, 1936

1. Object and Scope

The main purpose of the Payment of Wages Act, 1936, is to ensure regular and timely payment of wages to the employed persons, to prevent unauthorised deductions being made from wages and arbitrary fines being imposed being imposed on the employed persons. The Act extends to the whole of India.

2. Application of the Act

The Act applies in the first instance to the payment of wages of persons employed in any factory, to persons employed (otherwise than in a factory) upon any railway by railway administration, or either directly or through a sub-contractor by a person fulfilling a contract with the railway administration.

The Act also applies to persons employed in an "industrial or other establishment" specified in sub-clauses (*a*) to (*g*) of clause (*ii*) of Section 2.

3. Industrial or other establishments covered by the Act

According to section 2 (*ii*), the term 'industrial and other establishments' means any:

(*a*) Tramway service, or motor transport service engaged in carrying passengers or goods or both by road for hire or reward;

(*b*) Air transport service other than such service belonging to, or exclusively employed in the military, naval or air forces of the Union or the Civil Aviation Department of the Government of India;

(*c*) dock, wharf or jetty;

(*d*) inland vessel mechanically propelled;

(*e*) mine, quarry or oil-field;

(*f*) plantation;

(*g*) workshop or other establishment in which articles are produced, adapted or manufactured, with a view to their use, transport or sale;

(*h*) establishment in which any work relating to the construction development or maintenance of buildings, roads, bridges or canals or relating to operation connected with navigation, irrigation development or maintenance of buildings, roads, bridges or mission and distribution of electricity or any other form of power is being carried on;

(*i*) any other establishment of class of establishments which the Central Government or a State Government may, having regard to the nature thereof, the need for protection of persons employed there in and other relevant circumstances, specify, by notification in the Official Gazette.

4. Who are governed by the Act?

According to Section 1 (6), nothing in the Act shall apply to wages payable in respect of a wage period which, over such wage period, average one thousand and six hundred rupees a month or more. This means that only those employed persons who are normally drawing less than one thousand and six hundred rupees a month are converted by the Act.

5. Important Definitions

Employed person [Section 2 (*i*)]

"Employed person" includes the legal representative of a deceased employed person (This definition makes it possible for the legal representatives of a deceased employed person to prefer a claim relating to non-payment of wages or any unauthorised deductions therefore)

Employed [Section 2 (*ia*)]

"Employer" includes the legal representative of a deceased employer. After the death of employer his legal representative can be held liable for the payment of wages due to employed persons. The liability of the legal representative is limited to the extend of the value of the estate inherited by him.

Factory [Section 2 (*ib*)]

"Factory" means a factory as defined in clause (*m*) of Section 2 of the Factories Act, 1948 and includes any place to which the provisions of that Act have been applied under Sub-section (1) of Section 85 there of

Wages [Section 2 (*vi*)]

"Wages" means all remuneration (whether by way of salary, allowance or otherwise) expressed in terms of money or capable of being so expressed which would, if the terms of employment, express or implied were fulfilled, be payable to a person employed in respect of his employment or work done in such employment, and includes:

(*a*) any remuneration payable under any award or settlement between the parties or order of a Court;

(*b*) any remuneration to which the person is entitled in respect of overtime work or holidays or any leave period;

(*c*) Any additional remuneration payable under the terms of employment (whether called a bonus or by any other name);

(*d*) Any sum which by reason of the termination of employment of the person employed is payable under any law, contract or instrument which provides for the payment of such sum, whether with or without deductions, but does not provide for the time within which the payment is to be made;

(*e*) Any sum to which the person employed is entitled under any scheme framed under any law for the time being in force.

But does not include

1. Any bonus whether under a scheme of profit sharing or otherwise which does not form part of the remuneration payable under the terms of employment or which is not payable under any award or settlement between the parties or order of a Court;
2. The value of any house accommodation, or of the supply of light, water medical attendance or other amenity of any service excluded from the computation of wages by a general or special order of the State Government;
3. Any contribution paid by the employer to any pension or provident fund, and the interest which may have accrued thereof;
4. Any travelling allowance or the value of any travelling concession;
5. Any sum paid to the employed person to defray special expenses entailed on him by the nature of this employment; or
6. Any gratuity payable on the termination o employment in cases other than those specified in Sub-Clause (*d*).

Remuneration in terms of money or capable of being so expressed.

The remuneration payable as above may be in the nature of salary but all other allowances payable to an employed person in terms of his employment. Therefore D.A. granted in cash at particular rate becomes a part of wages payable to an employee under the contract of his employment.

❏❏❏

Annexure—IX

MINIMUM WAGES ACT, 1948

Introduction

The granting of minimum wages to the workers is one of the essential requirements. The wages should be sufficient to maintain the worker and his family. The object of the Act is to prevent exploitation of the workers. The minimum wages Act is a piece of social legislation intended to do social justice to the workers.

The Act extends to the whole of India. It came into force on 15th March, 1948. It applies to the employments which are listed in the schedule of the Act and in certain cases be extended to any other employment by the government.

Necessity of the Act

1. The workers are poorly organised and they cannot fight with the employer.
2. Trade union of workers are weak.
3. To fix the minimum wages and making it obligatory for the employer to pay to the workers.
4. By granting minimum wages, industrial peace is maintained.
5. To avoid the malpractice by the employer in payment of wages to the workers.

Important Terms of the Act

1. Appropriate Government

This means either the Central Government of the State Government, depending upon the type of employment.

2. Competent Authority

It means the authority appointed by the appropriate government by notification to ascertain from time to time the cost of living index number applicable to the employees in the scheduled employments.

3. Cost of Living Index Number

It means the index number ascertained and declared by the competent authority by notification in the official gazette to be the cost of living index number.

4. Wages

It means all remuneration capable of being expressed in terms of money be payable to a person employed in respect of his employment. It includes basic wages, dearness, bonus and house rent.

5. Minimum Wages

It represents the level below which wage cannot be allowed to drop. It should provide not nearly for existence, but for maintaining efficiency of the worker. Minimum wages should also provide for some measure of education, medical requirements and other amenities. The minimum wages are fixed on the cost of living index which is revised every six months effected from 1 April and 1 October.

The rates of mines workers and labourers employed in construction work in respect of the scheduled employment in the central sphere were revised vide Gazette Notification in January 2002. The national floor level minimum wage (introduced in 1996 in absence of a uniform national minimum wage) was last revised in September 2002.

Salient Features of Minimum Wages Act

1. Fixation of minimum wages.
2. Procedure for fixing and revising minimum wages.
3. Fixation of hours of work.
4. Methods of payments.
5. Maintenance of registers and records.
6. Inspection.
7. Penalties.

1. Fixation of Minimum Wages

The act provides the procedure for fixation of minimum wages for following workers:

(*i*) Different scheduled employments.
(*ii*) Different classes of work in the same scheduled employment.
(*iii*) Adults, children, apprentices.
(*iv*) Different localities.

Minimum rates of wages may be fixed by any one or more of the following wage period:

(*a*) By the hour
(*b*) By the day
(*c*) By the month
(*d*) By such other larger wage period as may be prescribe.

The following considerations will be taken into account in the fixation of the minimum wages

(*i*) The prevailing economic consideration.
(*ii*) The cost of living place.

(*iii*) The nature of the work to be performed, and

(*iv*) The conditions in which the work is to be performed.

2. Procedure for Fixing and Revising Minimum Wages

(*a*) Appointing the committees by the Government, which will hold enquiries and given the advice, in respect of fixation or revision of minimum wages.

(*b*) By notification in the official gazette publish its proposal for the information of the person likely to be affected thereby and specified a date, not less that two months from the date of notification, of which the proposal will be taken into consideration.

(*c*) On the information and advice the minimum wages are fixed or revised.

3. Fixation of Hours of Work

Where minimum wages in respect of any scheduled employment have been fixed, the appropriate Govt. may:

(*a*) Fix the number of hours for a normal working day inclusive of one or more specified intervals.

(*b*) Provide for a day of rest in every period of seven days.

(*c*) Provide for payment of remuneration in respect of such days of rest.

(*d*) Overtime shall be paid at the rate fixed under the act. These aregenerally double the normal rates.

4. Methods of Payments

(*a*) In a factory employing 1000 or less workers, the wages must be paid within seven days after the completion of the month.

(*b*) In a factory employing more than 1000 workers, the wages must be paid within ten days after the completion of the month.

5. Maintenance of Registers and Records

The employer is required to keep proper records and registers for payment of wages to each worker. The following details of each worker. The following details of each worker should be available:

(*a*) Rates of wages payable.

(*b*) The number of days in which overtime claimed by the worker.

(*c*) The deductions to be made from wages.

(*d*) The wages actually paid and date of payment.

(*e*) All other relevant details.

6. Inspection

The appropriate Govt. appoints inspectors for the enforcement of the Act. The inspector can enter in any premises and can inspect the documents of wages which he thinks fit for enforcing the rules, under the Act.

7. Penalties

The Act is administered by the State Govt. If any employer who pays less than the minimum rate of wages or violates any other provisions of the act, he can be punished with imprisonment upto six months or with a fine upto Rs. 500/- (five hundred) or both.

❑❑❑

Annexure—X

EMPLOYEE'S STATE INSURANCE ACT, 1948

Introduction

The object of this Act is to provide for certain benefits to employees in case of sickness, maternity and employment injury. The benefits which the Act wants to give to the employees are of a nature which would be difficult for any individual factory owner to provide for his own employees.

Application of the Act

This act extends to the whole of India. This Act is applicable to all the factories which are not seasonal, running with power and employing twenty of more persons. The workers drawing wages upto Rs. 6500/- are covered under this act with effect from 1 January, 1997.

The appropriate Govt. are authorised to extend the provisions of this Act partially or wholly to any establishment whether that may be industrial, commercial or agricultural.

The administration of this Act is done by an autonomous body the Employees' State Insurance Corporation which has local boards and medical councils for each region. The employees as well as employer give contributions to this scheme. The State and Central Govt. also finance the scheme. As on 31 December 2002, these were 140 ESI hospitals and 43 ESI annexes with 26,516 beds and 1451 dispensaries under the scheme covering about 72 lakhs employees.

Important Definitions

1. *'Appropriate Government'* means Central or State government.
2. *'Benefit Period'* means such period, being not less than twenty five but not exceeding twenty seven consecutive weeks or six consecutive months corresponding to the contribution period as may be specified in regulations.
3. *'Confinement'* means labour resulting in the issue of living child or labour after twenty six (26) weeks of pregnancy, resulting in the issue of a child whether alive or dead.
4. *'Contribution'* means the sum of money payable to the corporation by the employer in respect of any employee under the provisions of this Act.
5. *'Contribution period'* means such period being no less than twenty five but not exceeding twenty seven consecutive weeks or six consecutive months as may be specified in the regulations.

6. *'Corporation'* means the Employees' State Insurance Corporation set up under this Act.
7. *'Employment injury'* means a personal injury to an employee caused by an accident or an occupational disease arising in source of employment.
8. *'Exempted Employee'* means an employee who is not liable under this Act to pay the employees' contribution.
9. *'Seasonal factory'* means a factory which is exclusively engaged in one or more such processes as cotton ginning, sugar manufacturing etc.
10. *'Insured Person'* means any person who is or was an employee in respect of whom contributions are or were payable under this Act and, therefore, is entitled to the benefits provided in the Act.
11. *'Sickness'* means a condition which is of medical treatment and requires absentation from work on medical grounds.

Silent Feature of E.S.I. Act

1. Finance and Audit
2. Contribution
3. Benefits

1. Finance and Audit

(*a*) The act makes provisions for creation of fund called *'Employees' State Insurance Fund'*.

(*b*) The contributions towards fund are made by the employers and employees.

(*c*) The fund is held and administered by the corporation.

(*d*) The fund is utilized for payment of benefits and provisions of medical treatment, establishment and maintenance of E.S.I. Hospitals and dispensaries. The payment of fees, allowances, salaries are also paid from this fund to the employees of the corporation.

2. Contribution

According to the act employers and employees both make the contribution. The employer products the contribution of the employees from his salary. After additing his shares, the employer deposits the amount.

The E.S.I. corporation may appoint inspector to check the particulars amount of the fund submitted by the employer.

The rates of contribution depends upon the wages of the workers. The table shows the rates of the contribution the employees corresponding to average daily wages.

Group of employees whose average daily wages are	Employer's weekly contribution (recoverable from employees) Paise	Employer's weekly contribution Paise	Total weekly contribution Paise	Corresponding daily Standard benefit rate Paise
Below Rs 2/-	NIL	75	75	100
Rs. 2/- and above but below Rs. 3/-	40	80	120	130
Rs. 3/- and above but below Rs. 4/-	50	100	150	175
Rs. 4/- and above but below Rs. 6/-	70	140	210	250
Rs. 6/- and above but below Rs. 8/-	95	190	285	350
Rs. 8/- and above but below Rs. 12/-	125	250	375	500
Rs. 12/- and above but below Rs. 16/-	175	350	525	700
Rs. 16/- and above but below Rs. 24/-	275	550	825	1000
Rs. 24/- and above	375	750	1125	1500

3. Benefits

The Act provides the following Benefits:

1. Sickness benefit
2. Maternity benefit
3. Disablement benefit
4. Dependent benefit
5. Medical benefit

1. Sickness Benefit: It is in the form of periodical payment to any insured person in case of his sickness certified by

(*a*) A duly appointed medical practitioner.

(*b*) Any other person having qualification as prescribed in regulations.

A person entitled to claim sickness benefit shall receive benefit at the rates specified in schedule only for the period of sickness. The benefit cannot be given to any person for more than 56 days in any two consecutive benefit periods. No benefit is payable for the first two days except the medical attendance.

2. Maternity Benefit. This benefit consists of periodical payments to an insured woman in case of confinement, or miscarriage or sickness arising out of pregnancy, confinement, premature birth of child or miscarriage. Such woman should be certified to eligible for payments by a specified authority. Total maternity benefit is of 12 weeks of which not more than 6 weeks can precede the expected date of confinement.

3. *Disablement Benefit.* This benefit consists of periodical payment to an insured person suffering from disablement as a result of an employment injury.

During temporary disablement the amount of benefit is about half the average wages for the period of disablement. The disablement should not be less than three days.

For permanent disablement the injured persons are entitled to a pension in accordance with the proportionate loss in their earning capacity.

4. *Dependent Benefit.* This benefit becomes payable to the dependents of an insured person on his death as a result of an employment injury.

The daily rate of dependents benefits shall be 25% more than the standard benefit rounded to the next higher multiple of five paisa. It shall correspond to the average daily wages in the contribution period corresponding to the benefit period in which injury occurs. This rate is called the 'full rate'.

The benefit shall be payable to his widow and children as follows:

(*a*) To the widow during life or until remarriage an amount equivalent to 3/5 of the full rate.

(*b*) To each legitimate or adopted son, an amount equivalent to 2/5 of the full rate until he attains the age of 18 years.

The total benefit amount distributed to dependents does not exceed the full rate.

5. *Medical Benefits.* This benefit shall be given to an insured person or to a member of his family where the benefit is extended to his family.

This may be provided in the following form

(*a*) Out-patient treatment

(*b*) Attendance in a hospital, dispensary, clinic or other institution.

(*c*) By visits to the home of the insured person.

(*d*) Treatment as in-patient in a hospital or other institution.

6. *Funeral Benefit.* The benefit is payable to the eldest surviving member of the family or such person who actually incurs funeral expenses. This benefit shall not exceed Rs. 100/-.

❐❐❐

Annexure—XI

EMPLOYEE'S PROVIDENT FUND ACT, 1952

This Act provides some provisions for the future of industrial worker when he retires and for the benefits of his dependents in case of his death. The Government employees are given pension after retirement whereas industrial workers have the contributory provident fund scheme.

Application: This Act was passed in the year 1952 and extends to whole of India. This Act is applicable to

(*a*) Every establishment which is a factory in which 20 or more persons are employed.

(*b*) Any other establishment employing 20 or more persons which the central government may by notification specify in this behalf.

1. Contribution to the Fund

Retirement benefits in the form of provident fund, family pension and deposit linked insurance are available to employees under the Employees Provident Fund and MP Act 1952. As on 21 April 2001, the Act covered 180 industries/class of establishments. Coverage under the Act is presently restricted to establishments employing 20 or more persons. The present wage ceiling for applicability of the Act is Rs. 6500 with effect from 1 June 2001. The rate of contribution is 12% in respect of 175 industries/establishments employing 20 or more persons. Under the Act employers are required to make a matching contribution.

2. Protection Against Attachment

The amount standing to the credit of any member in the fund shall not in any way be capable of being assigned or charged. It shall not be liable to attachment under any decree or order of any court in respect of any debt or liability incurred by the employer. In case of the death of the employee, the amount of fund is paid to the dependents.

3. Reduction of Wages of the Employees

No employer will reduce directly or indirectly the wages of any employee due to contribution made by him (employer) towards the provident fund. The total quantum of benefits to the employees will not be reduced by the employer because of his liability for the payment of contribution to the fund.

4. Administration of the Act

Various boards have been formed and employees have been appointed by the Govt. for the administration of the Act.

(*a*) Central board

(*b*) State board

(*c*) Inspectors

(*d*) Appointment of Central Provident Fund Commissioner and Regional Provident Fund Commissioner.

Annexure—XII

INDUSTRIAL DISPUTE ACT, 1947

The whole history of labour struggle indicates the continuous demand for fair return to labour. Industrial disputes are mainly the result of dissatisfaction amongst the labour as regards their existing labour conditions. With the above in mind, the Industrial Dispute Act came into force in 1947 and aims at settling the industrial disputes on a new pattern known under the Act as adjudication machinery.

- This act aims at making industrial peace through voluntary negotiations and compulsory adjudication.

The act makes provision for settlement of industrial disputes between employees (workers) and employer.

Important Aspects of the Act

(*i*) An industrial dispute may be referred to an Industrial tribunal.

(*ii*) An award shall be binding on both the disputing parties for a period not exceeding one year.

(*iii*) Strikes and lockouts are prohibited during the pendency of,
- conciliation and adjudication proceedings;
- settlement reached in the course of conciliation proceedings,
- awards of industrial tribunals declared binding by the appropriate Government.

(*iv*) In public interest or emergency, appropriate government can declare the following industries to be a public utility service for a maximum period of 6 months:
- Transport by land, air, and water
- Foodstuff
- Coal
- Cotton textile, and
- Iron and Steel.

Important Definitions

1. *Award* means an interim or a final determination of an industrial dispute. Decisions of Labour Court and Industrial Tribunals are awards.
2. Average pay means the average of wages payable to workman :
   - in case of monthly paid workman, in the 3 calendar months;
   - in the case of weekly paid workman, in the 4 completed weeks; and

– in the case of daily paid workman, in the 12 full working days.

3. *Industry* means any business undertaking, manufacturer, etc. It includes any calling, service, employment, handicraft or industrial occupation or a vocation of workman.
4. *Industrial dispute* means
   - any dispute or difference,
   - between employers and employes, employers and workmen, workmen and workmen; and
   - connected with employment, or non-employment, or terms of employment or condition of labour, or any person.
5. *Lay-off* means
   - Failure, refusal or inability of an employer, on account of shortage of coal, power, raw material, accumulation of stock, breakdown of machinery or for any other reason, to continue to employ workers (in his industry) whose names are borne on the muster-rolls and who have been retrenched. Lay-off is a temporary phase; the employee-employer relation do not come to an end, but are simply suspended for some period (of emergency).
6. Lock-out means the closing of a place of employment or suspension of work or the refusal by an employer to continue to employ any number of workers employed by him.
7. *Public utility* services mean Railway or transport service, postal, telegraph or telephone service,

   Any industry supplying power, light water,

   Sanitation,

   Foodstuffs, and

   Coal, textile, etc.
8. *Retrenchment* means the termination of the services of a worker by the employer for any reason whatsoever other than due to disciplinary action. Retrenchment does not include,
   - Voluntary retirement of the worker,
   - Termination of services on reaching the age of superannuation, and
   - Termination of services on the ground of continued ill health.
9. *Strike* means refusal to work or cessation of work by a body of workmen for enforcement of a demand against the employer during and industrial dispute.
10. *Settlement*. It implies a settlement arrived at in the course of conciliation proceeding. It includes a written agreement between the workers and employer.
11. *Workmen* means any person (including an apprentice) employed in any industry for hire or award to do manual, technical, supervisory or clerical work.

Authorities Under this Act for Settlement of Industrial Disputes Works Committee

- Any industry, in which 100 or more workers are and have been employed on any day in the preceding 12 months, shall constitute a works committee.
- Works committee shall have representatives of workers and employer both. Workmen representatives will not be less than those of employer in number.
- Works committee shall promote measures for securing and preserving amity and good relations between the worker and employer. It will comment upon matters of their common interest and try to compose any material difference of opinion in respect of such matters.
- Works committee shall smooth away friction that might arise between the workers and the employer in day-to-day work.

Conciliation Officers

Conciliation officers will be appointed by any Appropriate Government and they shall mediate and promote the settlement of industrial disputes.

Board of Conciliation

Appropriate Government may constitute a Board of Conciliation to promote settlement of an industrial dispute.

- Board of conciliation will be headed by a chairman (an independent person) and shall have two or four other members which will be representatives of the parties to dispute. Both parties, *i.e.,* workers and employer shall have equal number of representatives.

Courts of Enquiry

- Appropriate Government may constitute a court of enquiry to look into any matter connected with industrial dispute.
- Court (of enquiry) may consist of one or more independent persons.
- The court shall inquire and submit a report ordinarily within six months from the commencement of inquiry.

Labour Court

- The appropriate government may constitute Labour Court consisting of one person only for adjudication of industrial disputes relating to any matter specified in II *Schedule.*
- The matters within the jurisdiction of Labour Court as per II *schedule* are :

1. The propriety or legality of any order passed by employer under the standing orders.
2. The application and interpretation of standing orders.
3. Illegality or otherwise of a strike or Lock-out.
4. Dismissal of workers including reinstatement or relief to workers wrongly dismissed.

5. Withdrawal of any concession or privilege.

Presiding officer of one man Labour Court shall be,

1. a judge of high court; or
2. a district judge who has worked for more than 3 years etc.

Industrial Tribunals

Appropriate Government may constitute one person Industrial Tribunals for the adjudication of Industrial Disputes relating to matters specified in,

1. Schedule II (refer to Labour Court for details of Schedule II),
2. Schedule Ill—Matters such as Wages,

   Compensatory and other allowances, Hours of work and responses, Holidays and leave with wages, Bonus, provident fund, gratuity. Rules of discipline, and Retrenchment of workers, etc.
3. Presiding officer of the tribunal shall have the same qualification as that of a Labour Court.

National Tribunals

- Central Government may constitute National Industrial Tribunals for the adjudication of industrial disputes which involve questions of national importance.
- The presiding officer of the National Tribunal shall be
  1. an independent person;
  2. less than 65 year of age; and
  3. an existing or retired judge of a high court, etc.

Prohibition of Adjudication by Other Tribunals

If any reference has been made to a National Tribunal, no Labour Court or Tribunal, shall have jurisdiction to adjudicate upon any matter which is under adjudication before the National Tribunal.

Strikes and Lock-Out

1. No employee of a public utility service shall go on strike

(*i*) Without giving notice of strike within six weeks before striking; or

(*ii*) within fourteen days of giving such notice; or

(*iii*) before the expiry of date or strike mentioned in notice; or

(*iv*) during the pendency of conciliation proceeding.

2. No employer of any public utility service shall lock-out

(*i*) without giving notice of lock-out within six weeks before locking out, or

(*ii*) before the expiry of date of lockout mentioned in the notice, or

(*iii*) during the pendency of conciliation proceeding.

The employer shall send information of strike or lock-out to the specified authority on the day on which it is declared.

3. Prohibition of financial aid to illegal strikes and lock-out

- A strike or lock-out is illegal if it is declared in contravention of points 1 and 2 above (*i.e.,* sections 22 and 23 of the act, respectively),
- No body shall knowingly expend any money in direct support of an illegal strike or lock-out.

Lay-off and Retrenchment

- For definition of Lay-off and retrenchment refer *Important Definitions* given earlier in this Act.

Right of Workmen laid-off for compensation

- A worker having more than one year of continuous service under an employer, if is laid off, shall get compensation equal to 50% of the total of his basic wages and dearness allowance, subject to the following limitation
- Provided that if during any period of 12 months, a workman is so laid-off for more than 45 days, no such compensation shall be payable in respect of any period of the lay-off after the expiry of the first 45 days, if there is an agreement to that effect between the workman and the employer, etc.

Workmen not entitled to compensation in certain cases

No compensation shall be paid to a worker laid-off under following conditions:

1. If he refuses to accept alternative employment involving same wages, offered to him in the same concern or in any other concern of the same employer.
2. If the lay-off is due to a strike, etc.

Conditions precedent to retrenchment

No worker who has a continuous service of more than one year shall be retrenched by the employer until:

1. The worker has been given one month's written notice stating the reason for retrenchment and the period of notice has expired or the worker has been paid in lieu of such notice, wage for the period of notice.
2. At the time of retrenchment, the worker has been paid compensation equivalent to 15 days average pay for every completed year of service.

Closing down the undertaking

- An employer who intends to close down an undertaking shall serve, atleast 60 days before the date on which he intends closure to become effective, a notice to Appropriate Government stating clearly the reasons for closing the undertaking.
- Compensation by reference to section 25-FFF of the Act is provided to workers who are in continuous service for not less than a year with the undertaking.

Penalties

- Workers who commence illegal strike shall be punishable with imperishable with imprisonment extending up to one month and a fine of up to Rs. 50, or both.
- An employer who commences an illegal lock-out shall be punishable with imprisonment up to one month or with a fine up to Rs. 1000 or both.
- The Act also specifies penalties of upto 6 months imprisonment or a fine of upto Rs. 1000 or both for:
  1. Instigating for illegal strike or lay-out.
  2. Giving financial aid to illegal strike or lock-out.
  3. Breach of settlement or award.
  4. Disclosing confidential information etc.

❑❑❑

Annexure—XIII

LIST OF INDUSTRIES RESERVED FOR THE PUBLIC SECTOR

Schedule I

List of Industries Reserved for the Public Sector

1. Arms and ammunition and allied items of defence equipment, defence aircraft and warships.
2. Atomic energy.
3. Coal and lignite.
4. Mineral oils.
5. Mining of iron ore, manganese ore, chrome ore, gypsum, sulphur, gold and diamond.
6. Mining of copper, lead, zinc, tin, molybdenum and wolfram.
7. The substances specified in the Schedule to the notification of the Government of India in the Department of Atomic Energy SO 212(E), dated the 15th March 1995, namely:
   (*a*) Uranium, its compounds and minerals/ores/concentrates containing uranium including tailings containing uranium.
   (*b*) Thorium, its compounds and minerals/concentrates containing thorium including monazite.
   (*c*) Zirconium, its alloys and compounds and minerals/concentrates including zircon.
   (*d*) Beryllium, its compounds and its minerals/concentrates including beryl but excluding beryllium windows used for medical X-rays machines and gamma/ray machines.
   (*e*) Lithium, its compounds and its minerals/concentrates including lepklolite.
   (*f*) Deuterium and its compounds.
   (*g*) Plutonium and its compounds.
   (*h*) Neptunium and its compounds.
   (*i*) Columbite and tantalite.
   (*j*) Titanium ores and concentrates (ilmenite, ratile, loucexene).

(*k*) Nuclear grade graphite.

(*l*) Tritium and its compounds.

8. Railway transport.

Ministry *of Industry, Department of Industrial Development*
Circular No. W(3l)92-LP, dated 26-3-1993

Under the Industrial Policy of July 1991, the following minerals bearing areas have been reserved for the public sector:

(*i*) Mining of iron ore, manganese ore, chrome ore, gypsum, sulphur, gold and diamond.

(*ii*) Mining of copper, lead, zinc, tin, molybdenum and wolfram.

The National Mineral Policy of 1990 has been reviewed by the Ministry of Mines in the light of the New Industrial Policy of July 1991. Under the National Mineral Policy of 1993, the Government have deleted all the 13 minerals which had earlier been reserved under the National Mineral Policy for exclusive exploitation by the public sector and has thrown them open for exploitation by the private sector.

Consequent upon the amendment of National Mineral Policy, it has been decided to delete the above-mentioned 13 minerals and mineral bearing areas from the list of items reserved for public sector under the Industrial Policy Statement of July 1991. Necessary notification for deletion of the above-mentioned minerals and minerals areas is being issued separately under the provisions of Industries (Development and Regulation) Act, 1951.

Annexure—XIV

LIST OF AGENCIES

CB	: Commerical Bank
CECD	: Central Excise and Customs Department
CIPET	: Central Institute of Plastic and Engineering Tools
CSIR	: Council of Scientific and Industrial Research
CCIE	: Chief Controller of Imports and Exports
CITD	: Central Institute of Tool Design
CIF	: Chief Inspector of Factories
DEP	: Directorate of Export Promotion
DGSD	: Director General of Supplies and Disposal
DDCA	: Directorate of Drugs Control Administration
DIC	: District Industries Centre
DGTD	: Director General of Technical Directorate
EB	: Electricity Board
EDI	: Entrepreneurship Development Institute of India
EDP	: Agencies Conducting Entrepreneurship Development Programme
EPC	: Export Promotion Council
GMD	: Geology and Mining Department
ICAR	: Indian Council of Agricultural Research
ICICI	: Industrial Credit and Investment Corporation of India Ltd.
ICMR	: Indian Council Medical Research
IDBI	: Industrial Development Bank of India
IIC	: Indian Investment Centre
IIFT	: Indian Institute of Foreign Trade
IFCI	: Industrial Finance Corporation of India
IPB	: Invention Promotion Board
ISI	: Indian Standard Institute (Burearu of Indian Standards)
IDCs	: Industrial Development Corporation (State level)
IICs	: Industrial Investment Corporation (State level)
IC	: Industries Commissioner
IEB	: Industrial Extension Bureau/State Level Co-ordinating Agencies
IRL	: Industrial Research Laboratory
LA	: Local Authorities, Municipal, Panchayat, etc.

LC	:	Leasing Companies
MID	:	Ministry of Industrial Development
MMTD	:	Mineral and Metal Trading Corporation
MDC	:	Mineral Development Corporation
NB	:	Nationalised Bank
NIESBUD	:	National Institute of Entrepreneurship and Small Business Development
NRDC	:	National Research and Development Corporation
NISIET	:	National Institute of Small Industry Extension Training
NIDC	:	National Institute Development Corporation
NSIC	:	National Small Industries Corporation
PTC	:	Polytechnology Transfer Centres
RIMCO	:	Rural Industries and Marketing Corporation
RC	:	Registrar of Companies
RT	:	Registrat of Trade Marks
SBI	:	State Bank of India
SIDO	:	Small Industries Development Organisation
SISI	:	Small Industries Service Institute
SC	:	Office of the Salt Commissioner
SPC	:	State Chemicals and Pharmaceuticals Organisation of India
SEC	:	State Export Corporation
SFC	:	State Finance Corporation
SIC	:	State Small Industries Corporation
STC	:	State Trading Corporation
TC	:	Textile Commissioner
TCO	:	Technical Consultancy Organisation (of State)
WPCB	:	Water Pollution Control Board

❑❑❑

Annexure—XV

THE CONSUMER PROTECTION ACT, 1986

Chapter 1: Preliminary

1. Short title, extent, commencement and application
2. Definitions
3. Act not in derogation of any other law

Chapter 2: Consumer Protection Councils

4. The Central Consumer Protection Council
5. Procedure for meeting of the Central Council
6. Obejcts of the Central Council
7. The State Consumer Protection Councils
8. Objects of the State Council

8A. The District Consumer Protection Council

Chapter 3: Consumer Disputes Redressal Agencies

9. Establishment of Consumer Disputes Redressal Agencies
10. Composition of the District Forum
11. Jurisdiction of the District Forum
12. Manner in which complaint shall be made
13. Procedure on admission of complaint
14. Finding of the District Forum
15. Appeal
16. Composition of the State Commission
17. Jurisdiction of the State Commission

17A. Transfer of cases

18. Procedure applicable to State Commissions
19. Appeals

19A. Hearing of Appeal

20. Composition of the National Commission
21. Jurisdiction of the National Commission
22. Power of and procedure applicable to the National Commission

22A. Power to set aside ex parte orders

22B. Transfer of cases

22C. Circuit Benches

22D. Vacancy in the Office of the President

23. Appeal
24. Finality of orders
24A. Limitation period
24B. Administrative Control
25. Enforcement of orders of the District Forum, the State Commission or National Commission
26. Dismissal of frivolous or vexatious complaints
27. Penalties
27A. Appeal against order passed under section 27

Chapter 4: Miscellaneous

28. Protection of action taken in good faith
28A. Service of notice, etc.
29. Power to remove difficulties.
29A. Vacancies or defects in appointment not to invalidate orders
30. Power to make rules
30A. Power of the National Commission to make regulations
31. Rules and regulations to be laid before each house of Parliament.

After including the amendments made vide the Consumer Protection (Amendment) Act, 2002 [62 of 2002] which was passed by Rajya Sabha on 11.4.2002, Lok Sabha on 30.7.2002 {with some amendments} and agains by Rajya Sabha on 22.11.2002 and the President of India gave assent on 17.22.2002 and the notification was issue on 18.22.2002. The provisions of the Act are being brought into force w.e.f. 15.3.2003.

REFERENCE BOOKS

1. Forbat, John,"Enterpreneur ship", (2008), New Age International Publishers, New Delhi.
2. Havinal, Veerbhadrappa, "Management and Entrepreneurship" (2010), New Age International Publishers, New Delhi.
3. Joseph L. Massie, "Essentials of Management" (2003), Prentice Hall of India Private Ltd., New Delhi.
4. Singal, R.K., "Entrepreneurship Development and Management, (2009) Kataria and Sons, New Delhi.
5. Prasanna Chandra, "Financial Management," (2011) Tata Mc Graw Hill, New Delhi.
6. Chhabra, T.N., "Entrepreneurship Development" (2011), Sun India Publication, New Delhi.
7. Badhai, B "Entrepreneurship Development Programme," (2010), Dhanpat Rai and Co., Delhi.
8. Singal, R.K. et al, "Project Management," (2011), S.K. Kataria and Sons, New Delhi.
9. Krishnamulti, S, "Small Scale Industries" (2005) Orient Publishing Company, New Delhi.
10. Desai, Vasant, "Dynamics of Entrepreneurship Development and Management" (2005), Himalaya Publishing House, Mumbai.
11. Khanka. S.S., "Entrepreneurship Development"(2011), S. Chand and Company, Ltd., New Delhi.
12. Gupta, C.B., et al," Entrepreneurship Development in India'' (2000), Sultan Chand and Sons, New Delhi.
13. A Hanabork for New Entrepreneurs (1986), Entreprenerușhip Development Institute of India, Ahmedabad.
14. Geoffrey, G, et al, "The Practice of Entrepreneurship" International Labour Office, Geneva.

❑❑❑

UNIVERSITY PAPERS WITH SOLUTION HINTS

UNIVERSITY PAPER WITH SOLUTION HINTS

B.Tech

Seventh Semester Theory Examination, 2011-12

ENTREPRENEURSHIP DEVELOPMENT

Time: 3 Hours *Total Marks: 100*

Note: Attempt all questions:

1. **Attempt any four parts :** **(5×4=20)**
   (*a*) "Entrepreneurship can be considered as the creator of economy." Comment.
   (*b*) What are the differences between an entrepreneur and a manager?
   (*c*) What do you mean by project planning? Discuss.
   (*d*) "Entrepreneurship are made, not born." Comment.
   (*e*) What is a business plan?
   (*f*) Discuss various types of risk involved in entrepreneurship.

2. (*a*) Explain the theories of Entrepreneurship. (10)
   (*b*) What are the different stages in starting a small-scale industry? (10

   OR

   (*a*) Discuss various government policies for small scale industry. (10)
   (*b*) Explain various types of small scale industries. Also explain their characteristics. (10)

3. (*a*) What is project report and how a good project is prepared? (10)
   (*b*) Write short notes on:
   (*i*) Net Present Value
   (*ii*) Internal Rate of Return (10)

   OR

   (*a*) Discuss various methods of demand forecasting. (10)
   (*b*) Write short notes on:
   (*i*) Assessment of viability.
   (*ii*) Method of collection of information. (10)

4. "An entrepreneurship is required to consider a large number of factors for managing a project successfully." Discuss this statement. How is the targeting and positioning of a product done? (20)

   OR

   Discuss various laws connecting entrepreneurship. (20)

5. Write short notes on any two : (10×2=20)
   (*a*) Cost of capital approach (*b*) Risk analysis
   (*c*) Policies in public enterprises (*d*) Types of Business Ownership.

SOLUTION HINTS

1. (*a*) Refer section 1.6.
   (*b*) Refer section 1.12.
   (*c*) Refer section 7.2.
   (*d*) Refer section 1.21 (15) and (16)
   (*e*) Refer section 3.3.
   (*f*) Refer section 7.8.2.

2. (*a*) Refer section 1.13.
   (*b*) Refer section 2.11.

   OR

   (*a*) Refer section 2.9.
   (*b*) Refer section 2.7 and section 2.6.

3. (*a*) Refer section 4.4.
   (*b*) (*i*) Refer section 4.7.5.
   (*ii*) Refer section 4.7.4.

   OR

   (*a*) Refer section 4.7.
   (*b*) (*i*) Refer section 4.5.
   (*ii*) Refer section 3.5.

4. Refer section 1.10, section 3.2 and section 3.6.

   OR

   Refer section 9.5 and section 9.11.

5. (*a*) Refer section 7.6.
   (*b*) Refer section 7.8.3.
   (*c*) Refer section 7.9.
   (*d*) Refer section 9.4.

B.Tech

Seventh Semester Theory Examination, 2012-13

ENTREPRENEURSHIP DEVELOPMENT

Time: 3 Hours *Total Marks: 100*

Note: *Attempt all questions. Each question carries equal marks. Draw neat sketch whenever required.*

1. Attempt any four parts of the following. Each part carries equal marks: **(5×4=20)**

(*a*) Stating missions and strategies for a new project, suggest its business plan.

(*b*) What are the stages through which a small sector entrepreneur has to go through while obtaining a loan from a financial institution?

(*c*) Explain the role of small scale industries in the national economy.

(*d*) Describe the traits found in successful entrepreneurs.

(*e*) Why are markets, money and motivation considered essential ingredients to small firm planning?

(*f*) What is entrepreneurship? Describe how risk taking is an important element of entrepreneurship.

2. Attempt any FOUR parts of the following. Each part carries equal marks:

(*a*) An investment of Rs. 1,00,000 yields cash flows for 5 years as Rs. 25,000, Rs. 40,000, Rs. 50,000 and Rs. 30,000. Calculate the cost benefit ratio, interest rate is 12%.

(*b*) What is the IRR for the above problem?

(*c*) For a project for which initial investment is Rs. 3,00,000 and future cash flows are Rs. 60,000, Rs. 1,00,000, Rs. 1,20,000 and Rs. 1,50,000 for next 4 years. What is its NPV?

(*d*) What are mission and strategies? How does SWOT analysis help in setting up strategy for small business?

(*e*) Successful entrepreneurs are said to have certain common characteristics. Explain them.

(*f*) Why are markets, money and motivation consideration essential ingredients to small business?

3. Attempt any TWO parts of the following. Each part carries equal marks:

(*a*) Give examples of assets, liabilities, net worth, revenues, expenses, direct material, direct labour and overhead.

(*b*) What are manufacturing overheads:? How are they allocated?

(*c*) Define Quality management. Total Quality revolves on some concepts, what are these concepts?

4. **Attempt any TWO parts of the following. Each part carries equal marks:**
   (*a*) What is the difference between capital funding and operation funding in the business? What are diffeent source of fund of small business?
   (*b*) Entrepreneurs planning a business will most likely be asked for cash flow statements. The statement would help you plan expenditure and receipts of cash. Can you describe the steps that are needed to make a cash flow statement?
   (*c*) Prepare a Balance Sheet and Profit and Loss Account for a company from the data given:

Cost of goods sold	Rs. 1,20,000.00
Sales	Rs. 2,00,000.00
Operating expenses	Rs. 75,000.00
Total Assets	Rs. 1,60,000.00
Total Liabilities	Rs. 70,000.00
Net Owner's Equity	Rs. 90,000.00

5. **Attempt any FOUR parts of the following. Each part carries equal marks:**
   (*a*) What is means by 'limited liability' in a limited partnership? What are the advantages of changing the ownership? What are the advantages of changing the ownership (partnership) to a company form?
   (*b*) Why is it important to be aware of the laws concerning labour management relations?
   (*c*) State the implications of the "Consumer Protection Act" on the business unit. What is the difference between a copy right and a patent?
   (*d*) Explain the term EBIT (earning before income tax). How it is determined in an income statement? What are excise duty, tangible property tax, capital gain tax and income tax?
   (*e*) What is depreciation and why it is charged? Why is called non cash flow?
   (*f*) What is the role of government financial institution in financing business? What tax benefits they offer to export oriented goods?

SOLUTION HINTS

1. (*a*) Business Plan:
   (*i*) Missions: Project identification #3.2 and 3.2.1
   (*ii*) Strategies: Criteria for selecting a project. #3.2.4
   (*iii*) Business Plan: Preliminary Project Report: #4.2, 4.2.1, 4.2.2
   (*b*) Loan Application:
   (*i*) Preliminary Project Report: #4.2
   (*ii*) Feasibility study: #4.3
   (*iii*) Detailed Project Report: #4.4
   (*iv*) Project Viability: #4.5

(*c*) Role of small scale Industries in the National Economy, See #2.5
(*d*) Trails of successful Entrepreneurs, See # 1.8
(*e*) Essential Ingrandient for small firm planning.
   (*i*) Markets: Demand Analysis. # 4.7
   (*ii*) Money Financing of Enter prize. # 3.4
   (*iii*) Motivation: Enterpreneurstrip. #1.5
(*f*) Entrepreneurship. # 1.2

2. (*a*) Example 4.3.
   (*b*) Example 4.3.
   (*c*) Example 4.4.
   (*d*) #7.10.
   (*e*) Common characteristic of Entrepreneurs.
      (*i*) Working profile. # 1.8.
      (*ii*) Qualities of an entrepreneur. # 1.7.
   (*f*) See Q. 1(*a*).

3. (*a*)
   (*i*) Assets #7.6.3
   (*ii*) Liabilities are obligations or debts to be paid. Examples are creditors, bills payable, overdraft etc.
   (*iii*) Net Worth is the difference between assets and liabilities.
   (*iv*) Revenues are incomes from sales of current assets during current year. The revenues include sales of goods and services by the enterprize, earnings from interest, rent, divident, etc.
   (*v*) Expenses are the costs incurred for operations during an accounting period. These include capital expenditure for acquisition of fixed assets like land, buildings machinery etc. and revenue expenditure for purchase of current assets and factory purchases.
   (*vi*) Direct material #5.9.3 (1)
   (*vii*) Direct labour #5.9.3 (2)
   (*viii*) Overheads #5.9.3 (4)

3. (*b*) #5.9.5, 5.9.5.1 and 5.9.5.2.
   (*c*) #6.4.3, 6.4.3.1

4. (*a*) # 3.4
   (*b*) # 8.3.4
   (*c*) Example 5.1.

5. (*a*) **(*i*) Limited Liabilities:** General partners cover unlimited liability irrespective of amount of contributed capital and take part in daily activities of the business. Limited partners cover limited liabilities to the extent of contributed shares.

(*ii*) Disadvantages of Partnership Firm

9.4.2. (*iv*)

(*iii*) Advantages of company

9.4.9 (*ii*)

(*b*) # 9.11.

(*c*) Appendix XV

(*d*) (*i*) Income Statement # 8.8.2

(*ii*) Taxes #9.7, 9.8, 9.9

(*e*) Depreciation #5.9.4.

(*f*) #10.11, #9.10, #9.12.

❏❏❏

B.Tech

Seventh Semester Theory Examination, 2013-14

ENTREPRENEURSHIP DEVELOPMENT

Time: 3 Hours *Total Marks: 100*

Note: Attempt **all** questions. Each question carries equal marks. Draw neat sketch wherever required.

1. **Attempt any four parts of the following :**
   (*a*) Who is an entrepreneur? What these small business owners contribute to the national economy?
   (*b*) Explain the growth of small scale industries in India.
   (*c*) How does the government help small sector units and why does it to?
   (*d*) What is entrepreneurial decision process? Explain the various stages involved in this process.
   (*e*) Explain the stages in starting a small scale industry. What is service industry?
   (*f*) Describe how developing a good organization is an important phase of the entrepreneurship process. What kind of business activites are found in small business?

2. **Attempt any four parts of the following :**
   (*a*) Explain Demand Analysis.
   (*b*) Explain what are the important factors do you consider before preparing a project report?
   (*c*) What are the phases involved in the Project Identification?
   (*d*) A medical clinic wishes to invest in new equipment? It will generate ₹ 70,000 in receipts for next 5 years. What is the present value of this cash flow?
   (*e*) An investment has an initial cost of ₹ 2,50,000. The cash flow of the invest for next 4 years are ₹ 1,00,000, ₹ 90,000, ₹ 80,000 and ₹ 70,000. What is its NPV at interest rate of 8%?
   (*f*) For the above problem (question No. 2(e)) suggest the Internal Rate of Return (IRR).

3. **Attempt any two parts of the following :**
   (*a*) How the selling price of a product is determined?
   (*b*) What are different types of inventory small manufacturing firms can hold? How can the entrepreneurship decide how much to order and when to order? What is inventory turnover?
   (*c*) Given the information shown below for Jacks Inc., show the accounting equation. Then present the balance sheet and income statement :

Building	$ 22,500	Cash	$ 10,000
Loans payable	2,500	Retained earnings	17,500
Supplies Expenses	5,000	Utility Expenses	750

Miscellaneous expenses	7,000	Accounts payable	5,000
Supplies on hand	3,750	Sales	25,000
Accounts receivable	12,500	Capital stock	61,500
Salaries	12,500	Land	37,500

4. **Attempt any two parts of the following :**

(*a*) What are different ways of financing a company's financial needs? What is called a leveraged company? How the cost of capital is calculated?

(*b*) Why a cash flow statement is needed? Explain by giving an example. How cash flows can be improved? Suggest some methods.

(*c*) What is Business risks and how to manage them? Several types of risks that develop into business losses, enlist at least five risks and suggest methods of dealing with it.

5. **Attempt any four parts of the following :**

(*a*) Can you visualize just what is involved in developing a successful business plan?

(*b*) What are the advantages and disadvantages of the partnership?

(*c*) How to form a company? What are the advantages of a company over partnership firm?

(*d*) How is legitimate business deductions defined? What is double taxation as related to a company?

(*e*) Name some of the Financial Institutions, who help in setting of small scale industry? What do they see before extending a loan to them?

(*f*) What are some of the areas of legal implications for a small company?

❑❑❑

INDEX

A...

Advantage of
- Holding Inventories, 162
- Effective Prodcution Planning and Control, 129
- Preliminary Project Report, 72
- Profit Planning, 185
- Customers, 158
- Entrepreneur, 158

Advertisement, 157
Agencies to be Contacted for Project Implementation, 46
Agitation, Strikes, Lockouts, Picketing and Gherao, 150
Aim and Objectives, 25
Air (Preventive and Control of Pollution) Act, 1981, 214, 271
Analysis of Site Selection, 68
Application Forms, 229
Applications, 195
Assessment of Economic Viability, 103
Assistance Schemes, 248
Audit of Accounts, 122

B...

Background, 39
Balance Sheet, 121
Bank of Baroda, 234
Bank of India, 234
Barriers to Entrepreneurship, 17
Basic Principles, 187
Benefit Cost Analysis, 89
Books to be Maintained, 101
Business Ownership, 201

C...

Canara Bank, 234
Capital
- Operations, 189
- Expenditure, 180
- Expenditures of an Entreprize, 170
- Structure, 60

Cash Flow Calculations, 188
Characteristics of
- Decision-Making, 106
- Various Types of Small Scale Industries, 28

Characteristics, 179
Classification of
- Accounts, 101
- Industries, 40

Closing, 168
Commerical Banks, 232
Communication, 195
Comparison of Different Forms of Business Ownership, 211
Components of
- Control System, 194
- Cost of Capital, 172

Comprehensive Policy Package, 2000, 43
Computation of Business Income, 216
Composition Scheme, 221
Concept of Entrepreneur, 4
Constraints in Implementation of Projects, 54
Construction of Financial Flow Graph, 192
Consumer Protection Act, 1986, 308
Contents of Detailed Project Report, 79
Control of
- Capital Cost, 119
- Financial Flows, 192
- Operating Costs , 118

Controlling, 194
Cooperative Organization Society, 208
Cost Benefit Analysis, 58
Cost Estimation Procedure, 109
Cost of Capital and Project Planning and Control, 173
Cost of Capital, 172
Criteria for Selecting a Particular Project, 53

D...

Data Collection, 62
Decision Making, 106

Deductions, 217
Definition
- Small Scale Industries, 22
- Entrepreneur and Entrepreneurship, 5
- Management, 125
Definition and Scope, 220
Demand Analysis, 87
Depreciation , 112
Detailed Project Report (DPR), 79
Different Incentive Schemes by Central and State Governments, 44
Directing and Controlling, 168
Discounted Cash Flow Method, 90
District Industry Centres (DICs), 230
Double Entry Book Keeping, 102

E...

Economic
- Evaluation, 175
- Package for the MSMEs Announced on 13th May, 2020, 44
- Parameters, 27
- Theory of Entrepreneurship, 19
Electricity Safety Act, 1910, 214, 262
Elements of
- Cost and Ladder of Cost , 110
- Project Formulation, 55
Eligible Concerns for Financial Assistance, 247
Employee's
- Provident Fund Act, 1952, 296
- State Insurance Act, 1948, 224, 292
- Provident Fund Act, 1952, 225
Entrepreneur and Manager, 17
Excise Duty, 218
Executing, 168
Expected Costs, 108
Exploring Opportunities, 52
Export Incentives, 225
External Causes, 32

F...

Factor Affecting the Site Selection of Plant, 64
Factors Affecting Cost of Capital, 172
Factors Leading to Uncertainty, 176
Factory Act, 1948, 211, 253
Feasibility Analysis, 56
Field Study, 62
Field-study and Collection of Information, 62
Financial
- Analysis, 58
- Functions, 169
- Resources, 171
Financing of
- Project, 173
- Enterprize, 59
Formation of Business Organization, 200
Functions and Policies, 244
Functions of
- Advertising, 157
- Entrepreneur, 16
- Cost Estimation , 108
- Management, 126
- Production Planning and Control, 130

G...

Goods and Service Tax (GST), 220
Government Policy for Small-Scale Industries, 38
Grindlays Bank Limited, 235
Groups of Accounts, 102
Growth of Small Scale Industries in Developing Countries, 24

H...

Highlights of Industrial and Licencing Policy of India, 39
Hoselitz Sociological Theory, 18

I...

Importance of Project Identification, 54
Important Institutions at State Level, 238
Incentives and Tax Benefits for Export, 225
Incentives, 160
Income Tax, 215
Incorporation and Purpose, 243, 245, 248
Indian Bank, 234
Indian Boiler Act, 1923, 214, 259
Indian Experiment, 182

Indian Partnership Act, 1932, 277
Industrial Acts, 211
Industrial Development Bank of India, 245
Industrial Dispute Act, 1947, 149, 225, 298
Industrial Finance Corporation of India, 243
Industrial Policy Statements After Independence, 40
Industrial Relations, 147
Input Analysis, 58
Inspection, 135
Institutional Measures, 39
Internal Causes, 31
Internal Rate of Return Method, 92
Inventory Control, 161

J...

Job Satisfaction as Motivator, 161
Joint Hindu Family Business, 205
Joint Stock Company, 206

L...

Labour Laws, 223
Laws for Entrepreneur, 199
List of Agencies, 306
List of Industries Reserved for the Public Sector, 304

M...

Management of Working Capital, 189
Management, 243
Manpower or Human Resources, 171
Marketing Activities, 142
Marketing, 142
Material Balance and Output Method, 88
Material Purchase, 153
Max Weber's Theory of Social Change, 18
McClelland's Psychological Theory, 19
Meaning of Entrepreneur and Entrepreneurship, 4
Media for Advertisement, 158
Micro, Small and Medium Enterprizes Development Act, 2006, 43
Minimum Wages Act, 1948, 224, 288
Models of Inventory Management, 162

N...

National Bank for Agriculture and Rural Development (NABARD), 239
National Need, 6
National Small Industries Corporation, 236
Necessity of Management, 128
Need for
- Entrepreneurship Development, 6
- Finance, 59
- Inventory Management, 161
- Technical Entrepreneurs, 7
- Small Scale Industries, 24

Net Present Value Method (NPV), 92
New Classification of Types of Small Scale Industries, 24
New Entrepreneur Development Agency, 1986, 235
New Promotional Schemes, 244
New Small Enterprize Policy, 1991 After Liberalization of Economy, 42

O...

Objectives and Functions, 245, 247
Observations, 193
Office of Accountant, 101
Organizing, 168
Other Banks, 235
Other Types of Entrepreneurs, 15
Overhead Expenses, 116

P...

Package of Assistance, 247
Partnership Firm, 203
Partnership Laws, 201
Payment of Wages Act, 1936, 224, 285
Performance of Small Scale Enterprises, 27
Physical Resources, 170
Planning and Production Control, 128
Planning Cash Flows, 186
Planning, 167
Policies, 181
Post-tax Principle, 188
Practices, 181
Precautions, 59

Preliminary Project Report, 71
Preparation of
- Financial Reports, 120
- Preliminary Project Report, 72
- Project Report, 80
Principles of Control Process, 194
Principles of Management, 125
Principles of Taxation, 215
Problems Faced by Small Enterprizes, 33
Procedure of Decision-Making, 107
Procedure, 100
Process of Selection, 52
Procurement of Assets, 174
Profit and Loss Account, 121
Profit Planning and Programming, 184
Profit Programming, 185
Project
- Appraisal, 85
- Design and Network Analysis, 58
- Finance, 244
- Formulation, 55
- Identification, 51
- Life, 187
- Planning, 167
- Viability, 82
Promotional Measures, 38
Promotions as Motivator, 160
Protective Measures, 38
Public Enterprises, 179
Punjab National Bank (PNB), 233
Purchase Documents , 155
Purpose of Accountancy, 99

Q...

Qualities of an Accountant, 101
Qualities of an Entrepreneur, 10
Quality Control, 135

R...

Raising of Loans, 61
Reasons of Sickness in Small Industries, 30
Rehabilitation of Sick Units, 36
Relation Between Small and Large Units, 26
Registration Process, 24
Registration and Threshold, 221
Requirement of a Good Advertisement, 158
Requirements, 99
Rewards and Penalties for an Entrepreneur, 8
Risk Analysis, 175
Role of an Entrepreneur in Socio-economic Development, 9
Role of Government, 26
Role of Small Scale Industries in the National Economy, 27

S...

Sales Promotion, 152
Sales Tax, 219
Schumpeter's Innovation Theory, 19
Scope of Entrepreneurship, 7
Scope, 26
Selection of Business Organization, 200
Settlement of Industrial Disputes, 151
Small Industries Development Bank of India (SIDBI), 237
Small Industries Service Institutes, 236
Sole Proprietorship, 202
Sources for Identifying Business Opportunities, 52
Sources of Finance, 60
Sources of Funds, 244, 246, 248
Sources of Information, 227
Special Financial Services Provided by Institutes, 241
Special Incentives and Subsidy are given to Small-scale Sector
Special Incentives for Small-Scale Sector, 43
Special Schemes, 245, 247
Stages in Starting a Small-Scale Industry, 45
State Bank of India, 233
State Financial Corporations, 246
Stationery for Accounts, 101
Statistical Quality Control, 138
Steps in Enterprize Locations, 66
Stock-taking of Stores, 122
Study of Accounts and Stores, 122
SWOT Analysis, 182

Symptoms of Bad Management, 128
Syndicate Bank, 235

T...

Tax Benefits, 225
Taxation Benefits to Small-scale Industry, 222
Tax-payer's Obligation, 218
Techniques of Risk Analysis, 177
Techno-Economic
- Feasibility Study, 76
- Study, 57
Technological Resources, 171
Term Lending, 241
Theories of
- Entrepreneurship, 18
- Industrial Location, 64
Theory of Social Behaviour, 19
Time Schedule Plan for Starting a Business, 46
Total Quality Management, 140
Trading Account, 121
Trait Theory of Entrepreneurship, 18
Types of
- Small Scale Industries, 30
- Activities, 244
- Decision, 106
- Entrepreneurs by Clarence Danhof, 13
- Entrepreneurs, 13
- Financing Options, 173
- Project Risks, 176

U...

Underwriting of Capital Issues, 242
Unit Trust of India (UTI), 248

W...

Wages and Incentives, 159
Wages, 159
Water (Preventive and Control of Pollution) Act, 1974, 214, 267
Working Profile of an Entrepreneur, 12
Workman Compensation Act, 1923, 224, 280

Win Prizes

ATTENTION READERS

Win Prizes

We request for your frank assessment regarding various aspects of the book as given below :

ENTREPRENEURSHIP DEVELOPMENT

By: Dr. R.K. Singal

Please Fill up the Sheet in Neat and Clean Writing

(i) What topics of your syllabus, important for your examination point of view, are not covered in this book ?

....

....

....

....

(ii) What are the chapters wherein the treatment of the subject matter is not systematic or organised or updated ?

....

....

....

....

(iii) Have you come across misprints/mistakes/factual inaccuracies in the book? Please specify the chapters and the page numbers.

....

....

....

....

(iv) What other books on the same subject have you found/heard better than the present book? Please specify in terms of price and quality.

....

....

....

....

(v) Further suggestions and comments for the improvement of this book.

....

....

....

....

(PS: If need be please attach a separate sheet to write your views)

OTHER DETAILS

(i) Who recommended you this book ? (Please tick (✓) in the box near the option relevant to you.

☐ Teacher ☐ Friends ☐ Bookseller

....

....

....

....

(ii) Name of the recommending teacher, his designation and address :

....

....

....

....

(iii) Name and address of the bookseller you purchased the book from :

....

....

....

....

(iv) Name and address of your institution (Please mention the University or Board as the case may be)

....

....

....

....

(v) Your name and complete postal address :

....

....

....

....

(vi) Course you are enrolled in :

....

....

....

....

The best assessment will be awarded each month: The award will be in the form of our publications as decided by the Editorial Board.

Please mail the filled up coupon at your earliest to :

S.K. KATARIA & SONS®

4885/109, Prakash Mahal, Dr. Subhash Bhargav Lane,
Opposite Delhi Medical Association, Daryaganj, New Delhi–110002 (INDIA)
Phone: +91-11-43551243, Mobile: +91-9871775858, 7678425209
e-mail: katariabook@gmail.com, katariabooks@yahoo.com
Website: www.skkatariaandsons.com